LIGHT A

DARK

LIGHT AND DARK

DAVID GREENE

I*o*P

Institute of Physics Publishing
Bristol and Philadelphia

British Library Cataloguing-in-Publication Data

A catalogue record for this book is available from the British Library.

ISBN 0 7503 0874 5

Library of Congress Cataloging-in-Publication Data are available

Commissioning Editor: Nicki Dennis
Production Editor: Simon Laurenson
Production Control: Sarah Plenty
Cover Design: Frédérique Swist
Marketing: Nicola Newey and Verity Cooke

Published by Institute of Physics Publishing, wholly owned by The Institute of Physics, London

Institute of Physics Publishing, Dirac House, Temple Back, Bristol BS1 6BE, UK

US Office: Institute of Physics Publishing, The Public Ledger Building, Suite 929, 150 South Independence Mall West, Philadelphia, PA 19106, USA

Typeset in LaTeX 2_ε by Text 2 Text, Torquay, Devon
Printed in the UK by J W Arrowsmith Ltd, Bristol

CONTENTS

PREFACE   ix

1   ESSENTIAL, USEFUL AND FRIVOLOUS LIGHT   1
   1.1   Light for life   1
   1.2   Wonder and worship   4
   1.3   Artificial illumination   6
        1.3.1   Light from combustion   6
        1.3.2   Arc lamps and filament lamps   9
        1.3.3   Gas discharge lamps   13
   1.4   Light in art and entertainment   15

2   PATTERNS OF SUNLIGHT   19
   2.1   The year   19
   2.2   Equinoxes and eccentricity   22
   2.3   The length of a day   26
   2.4   The length of daylight   31
   2.5   The length of a second   37

3   MONTHS AND MOONLIGHT   41
   3.1   The lunar month and the lunar orbit   41
   3.2   The lunar nodes and their rotation   43
   3.3   The lunar day   48
   3.4   The length of moonlight   50
   3.5   Eclipses and Saros cycles   55
        3.5.1   Eclipses and history   55
        3.5.2   The Saros cycle   55
        3.5.3   Total and annular solar eclipses   61
   3.6   Tides   63

4   HISTORY, DATES AND TIMES   67
   4.1   Solar calendars   67
   4.2   The Roman Catholic Church and the development
        of astronomy   70

4.3	The start of the year	72
4.4	Lunar and other calendars	73
4.5	Time zones	77
4.6	The International Date Line	80

5 LIGHT AND THE ATMOSPHERE — 84
5.1	Scattered light and twilight	84
5.2	Polarization of light	88
5.3	Rainbows	94
5.4	Cloudy skies	98
5.5	Halos	99

6 SEEING THE LIGHT — 103
6.1	The human eye	103
6.2	Colour vision and colour blindness	109
6.2.1	Colour vision	109
6.2.2	Colour blindness	112
6.3	Polarization sensitivity	114
6.4	Speed of response	115
6.5	Optical illusions	118

7 ZOOLOGICAL DIVERSIONS — 129
7.1	Colour vision in animals	129
7.2	Zebras	131
7.2.1	Species and subspecies	131
7.2.2	Other zebra-striped animals	133
7.3	Piebald coats and unusual goats	134
7.4	Jellicle cats are black and white	138
7.5	Cephalopods	142
7.6	Lighting up for a mate or a meal	144
7.6.1	Bioluminescence in insects	144
7.6.2	Bioluminescence in deep-sea fish	146
7.7	More anatomical oddities	147

8 INFORMATION IN LIGHT — 150
8.1	Lighthouses	150
8.2	Semaphores for optical telegraphy	154
8.3	Morse, Mance and the heliograph	163
8.4	Bell and the photophone	166

Contents

9   LIGHT IN THE ERA OF ELECTRONICS   170
    9.1   Electronics 1900–1960   170
          9.1.1   Early rectification devices   170
          9.1.2   The solid-state rectifier   172
          9.1.3   The transistor   175
    9.2   New semiconductors for optoelectronics   177
    9.3   Optoelectronic semiconductor devices   182
    9.4   Bright light from cool solids   187

10 OPTICAL COMMUNICATION TODAY   193
    10.1 Waveguides and optical fibres   193
    10.2 The transparency of glass   195
    10.3 Optical fibres   198
    10.4 Optical amplification   203
    10.5 Conveying sound by light   204
    10.6 The long and the short of optical communication   210

    BIBLIOGRAPHY   213

    INDEX   215

PREFACE

In December 2001 Martin Creed was awarded the Turner Prize worth £20 000 for a work of contemporary art entitled 'The Lights Going On and Off'. It consists of an empty room with its most conspicuous feature aptly described by its title. Clearly a book on a similar theme is timely, though unlikely to be so financially rewarding.

This book brings together a wide range of topics that would normally be found in separate texts classified as astronomy, zoology, technology, history, art or physics. The connection is the theme of light and dark, which may alternate either in time or in space. In the time domain, slow variations often determine when animals mate and sleep, patterns defined in seconds provide navigational information for sailors and flashes of almost incomprehensible brevity convey messages and data around the world. Spatial patterns in black and white define the area on which chess players compete and enable the computer at the supermarket checkout to distinguish baked beans from jam tarts.

This book is intended to provide entertainment as well as instruction, and is in no way a comprehensive textbook for formal courses. For some more detailed accounts of particular topics you should refer to the suggestions for further reading. I have also mentioned places where you can look at such things as light-emitting fish and military heliographs. I have carefully avoided any mathematical analysis, but assume that readers will not be terrified by information presented in diagrams and graphs. Some parts of the book may be useful to students reluctantly following a science course to meet requirements for a broad curriculum. The topics reflect some quirks of my own personality and history, but generally they have been chosen because they are not far from the experience of most readers. There is information that could lead to more rational choices when buying sunglasses or light bulbs.

Subjects like photonic crystals and polaritons are deemed too abstruse for inclusion, even though they are fascinating topics for scientists currently investigating interactions between light and matter.

Although zebras and rainbows are familiar, they have interesting features that frequently pass unnoticed. An aim of this book is to encourage the reader to look more carefully at such sights. Many people are aware of the spring and autumn equinoxes but do not realize that the average time between sunrise and sunset is about twelve and a quarter hours in Britain on 21st March and 21st September. I hope that readers will not only take notice of this apparent paradox but also understand the reasons. I have included some easily demonstrated visual effects that were first noted in the 19th century but are rarely included in science courses. The human eye perceives colours in certain moving black-and-white patterns and has some ability to identify polarized light.

Patterns of light and dark are not always natural phenomena to be observed and enjoyed. Human ingenuity allows them to be created for entertainment or for conveying information. For thousands of years light has been a carrier of messages, often for military and naval purposes. In the 19th century army signallers used sunlight to send messages in less than a minute across distances that took a horseman a day. By the middle of the 20th century, copper wires and radio waves seemed to have captured most of the market in rapid long-distance communication. Nevertheless fifty years later incredibly short flashes of infrared light convey huge amounts of data and speech from continent to continent at an extraordinarily low cost.

It is not essential to read all the chapters in strict numerical order, but some of them do require acquaintance with earlier material. Chapter 1 makes no great demands on the reader. Chapters 2 and 3 are concerned with astronomical cycles involving the Sun and the Moon and form a basis for understanding the calendars described in chapter 4. Chapters 6 and 7 are mainly biological and do not require any knowledge of the contents of chapters 2, 3 and 4. Readers seeking information about vision and light emission in the animal kingdom and already familiar with polarized light could also miss out chapter 5, which is about light in the sky. The use of light in human communications is described in chapters 8 and 10. Chapter 9 provides some technical and historical background needed to appreciate the modern optical communication systems described in chapter 10. These last three chapters are best read in sequence but could be tackled without reading any of the first seven.

Numerous people have made helpful inputs during the writing of this book. Nick Lovibond at the Australian Antarctic Division, Michael Land at the University of Sussex and S Krebs at the Schweizerischer Ziegenzuchtverband kindly supplied information to a total stranger. Clare McFarlane, Steve Oliver and Sue Wheeler commented constructively on various chapters. Jane Greene produced some of the drawings and read the whole text critically more than once. For the final careful review of the entire book and many improvements, both literary and technical, thanks are due to Graham Saxby.

Figures 1.6, 5.12, 5.16 and 7.10 are reproduced by kind permission of the National Gallery, Clare McFarlane, Kip Ladage and Oxford Scientific Film, respectively.

David Greene
Harlow
March 2002

1

ESSENTIAL, USEFUL AND FRIVOLOUS LIGHT

1.1 Light for life

Life on Earth is almost totally dependent on the regular input of energy that is supplied by radiation from the Sun. The input maintains the temperature of most of the sea and the land surface within a range that allows living creatures to function. Some of the sunlight provides the energy for photosynthesis, the process plants use to convert carbon dioxide and water into oxygen and carbohydrates such as glucose. The products of photosynthesis contain more energy than the starting materials, and other life forms, such as animals and fungi, can exploit the stored energy. The animals inhale the oxygen and consume the plants, either directly (herbivores) or indirectly (carnivores), and return carbon dioxide and water to the environment.

Photosynthesis is a multistage chemical process in which the key role is played by chlorophyll. There are several subtly different forms of this complex organic compound, but the molecules of all forms contain just one atom of magnesium. Chlorophyll obtains the energy necessary for the synthesis of carbohydrates by selectively absorbing light from both ends of the visible spectrum, as shown in figure 1.1. Light in the middle of the visible spectrum is not absorbed but reflected, so the leaves of most plants appear green. Whereas colours within the visible spectrum have wavelengths from 370 to 740 nanometres (nm) (1 nm $= 10^{-9}$ m) and are either beneficial or harmless, ultraviolet light has a destructive

1

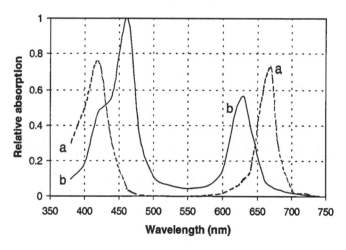

Figure 1.1. Absorption spectra of chlorophyll. Chlorophyll a and chlorophyll b have almost identical molecular structures, each with one magnesium, four nitrogen and fifty-five carbon atoms, but a has two more hydrogen atoms and one fewer oxygen atom than b. Leaves contain chlorophylls and appear green because absorption is least for wavelengths in the middle of the visible spectrum. The high absorption at longer and shorter wavelengths provides the leaves with the energy needed for photosynthesis.

effect on living cells, particularly when the wavelength is below 300 nm.

The relative amounts of the various colours or wavelengths of light emitted by a hot body depend on its surface temperature. The law that describes the relationship between the temperature of a hot emitter and the intensities of the emitted radiation at various wavelengths was discovered and explained about a hundred years ago by Max Karl Ernst Ludwig Planck, who was a professor of physics in Berlin. The explanation involved a new concept known as quantum theory, a major advance in physics for which Planck received a Nobel Prize in 1918. The precise mathematical form of the relationship between emitter temperature and the emitted radiation is a little too complex for presentation here, but some of its consequences are illustrated in figure 1.2.

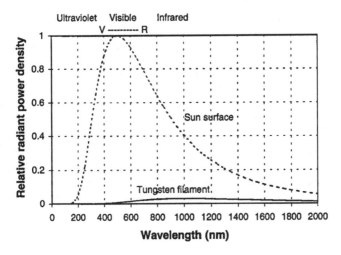

Figure 1.2. Light output from the Sun and from a tungsten filament lamp. The temperature of the surface of the Sun is around 5800 K or 5500 °C and so the maximum intensity of the emitted light lies within the wavelength range detectable by the human eye. The tungsten filament in an ordinary lamp bulb has a working temperature of about half that of the Sun's surface. Consequently the intensity of the emission is much lower, and more than 90 per cent of it is in the infrared.

This figure compares the light emitted at different temperatures by a material that would appear black at normal temperatures. In one case the material is at the temperature of the surface of the Sun and in the other case at half that temperature, which is reached by the tungsten filament in an ordinary light bulb. It can be seen that doubling the temperature increases the greatest intensity by a factor of 2^5 or thirty-two and halves the wavelength at which the peak occurs. The figure also shows that the most intense radiation from the Sun is in the visible part of the spectrum and that there is a large rise in intensity in the ultraviolet as the wavelength increases from 200 to 400 nm. Although ozone in the stratosphere at heights between 18 and 35 km absorbs much of the ultraviolet light with wavelengths between 200 and 350 nm, the temperature of the surface of the Sun is of critical importance for life on Earth. If the Sun were slightly hotter, its output would contain a lethal proportion of ultraviolet light. If the Sun were cooler, the output of blue light might be insufficient for photosynthesis by chlorophyll to proceed at an adequate rate.

Nevertheless there are hundreds of species that derive the energy to sustain life without directly or indirectly relying on photosynthesis. Deep in some oceans where no sunlight reaches, there are volcanic vents that release heat and sulphur compounds into the water. Here live bacterial colonies that base their metabolism on the available materials and energy sources. In turn, other living organisms such as tube worms exploit these bacteria. Because these worms do not have a gut through which food passes, it is almost certain that they do not benefit from a food chain beginning near the surface.

1.2 Wonder and worship

Some underground-dwelling creatures, such as earthworms and naked mole rats, have no functional eyes. To them it is immaterial whether it is day or night, summer or winter. In contrast, most other animals are strongly influenced by daily and annual variations in the amount of light. *Homo sapiens* is affected and also intrigued by the patterns of light and dark, which feature in human thoughts about art, religion and science. Furthermore, our species has developed an impressive ability to create artificial light for its own purposes, both practical and recreational. Other animals are able to create light, but we shall leave that topic until chapter 7, and concentrate our attention on humans.

Ancient artefacts indicate the importance of the natural cycles of light and darkness in the lives and thoughts of people living thousands of years ago. Some prehistoric communities devoted a substantial fraction of their effort to structures designed with the positions of the Sun in mind. Ireland can boast of a massive example from the Neolithic era. Radiocarbon dating indicates that the passage tomb at Newgrange in County Meath was constructed around 3200 BC, some 600 years before the building of the Great Pyramid of Cheops in Egypt. This circular structure is about 85 metres in diameter and has a slightly convex upper surface about 10 metres high at the centre. From an entrance in the near-vertical exterior wall, a passage about 18 metres long and

1 metre wide leads into the main chamber, which has a ceiling almost 6 metres high. The passage is aligned in the direction of the rising Sun at the winter solstice, and a hole above the entrance allows light from the rising Sun to penetrate to the far wall of the main chamber at this time.

In Great Britain, Stonehenge is the best-known site with an obvious alignment to a direction of astronomical significance. Development of the site is thought to have begun before 3000 BC, but the assembly of large stones occurred around a thousand years later. The bluestones from southwest Wales probably arrived around 2150 BC, some 150 years before the huge sarsen stones which provide Stonehenge with its most obvious and memorable characteristics. The major axis is aligned to the sunrise at the summer solstice and the sunset at the winter solstice. In the 1960s it was proposed that Stonehenge had also been used to observe and record lunar cycles. The availability of data about the directions of both Sun and Moon might have permitted the prediction of eclipses, but the majority of archaeologists are sceptical about such hypotheses.

Light features prominently in ancient religious texts. At the beginning of the Old Testament is the Book of Genesis, which has been estimated to date from the 8th century BC. The first five verses of the first chapter mention darkness, light, night and day. The 14th to 19th verses are concerned with seasons and years, the Sun, the Moon and the stars.

Light acquired a metaphorical as well as a physical significance. 'Enlighten' means 'inform' or to 'provide understanding', particularly in a religious context. The name 'Buddha' means 'enlightened one', and adherents of eastern philosophies and religions such as Hinduism or Buddhism strive towards a state described as enlightenment. Deities linked to the Sun have been widespread, from the Aztecs to the Egyptians. According to the Gospel of St John, Jesus claimed to be 'the light of the world'. The ancient Greeks ascribed the westward movement of the Sun to the deity Helios, who drove across the sky in a chariot pulled by four horses. Each night, he sailed back on a mythical sea to the start of his daily run. The westward movement of the Moon was associated with his sister, the goddess Selene, whose chariot was drawn by only two horses. Although the apparent speed of the Moon across the sky is slightly less than that of the Sun, the difference in speeds does not appear to justify a power ratio of two to one. It is extremely unlikely that the ancient Greeks were considering the relative masses of the Moon and the Sun, so the lower horse-power of the celestial vehicle with the female driver may simply have arisen from hypothetical differences in the characteristics of male and female divinities. Nevertheless there is no worldwide agreement about the genders of the Sun and the Moon, either in characteristics or in grammar. Arabs perceive the Sun to be feminine and the Moon to be masculine. In French the Sun is *le soleil* and the Moon is *la lune*. In German the genders are reversed, the Sun being *die Sonne* and the Moon *der Mond*.

1.3   Artificial illumination

1.3.1   Light from combustion

In both the practicalities of daily existence and the attempts to understand life's significance and meaning, light has been very important for *Homo sapiens*. It is therefore not surprising that humans devoted considerable amounts of thought and resources to achieving creation and control of this precious but fleeting commodity. The ability to generate light has existed for more than 12 000 years. Cave walls have been found with pictures of animals painted by Palaeolithic artists. In some locations, such as Niaux on the French side of the Pyrenees, the paintings are hundreds of metres from the cave entrance, and must have needed a fairly reliable source of artificial light for both creation and viewing. The light probably came from burning animal fat such as tallow, held in a bowl and drawn up a wick made from vegetable fibres.

Man-made light sources for religious rituals are mentioned in the Old Testament books of Exodus (chapters 25 and 37) and Numbers (chapters 4 and 8). The early history of candles is hard to trace, but it is clear that well before 500 BC they were being used by several communities around the Mediterranean, including the Etruscans and the Egyptians. Their function was not merely to provide light to prolong the time available for productive work, but to keep evil spirits away. In the fourth act of Shakespeare's *Julius Caesar*, Brutus remarks 'How ill this taper burns' as the ghost of Caesar appears before him. In the 16th century Shakespeare's plays were performed in daylight at the Globe

Theatre. To make the appearance of the ghost plausible, the audience needed to be told explicitly about the dimming of the light in Brutus's tent.

Unfortunately burning animal fat produces an unpleasant odour. Consequently, religious ceremonies often demanded less smelly but more expensive candles made from beeswax, a material secreted by glands on the abdomens of bees, which use it for making honeycombs. Beeswax candles are still popular today because of their pleasant aroma. In the 19th century new materials for making candles became available, including spermaceti wax from whales and stearic acid obtained by the breakdown of animal fat. With the development of the petrochemical industry in the 20th century, paraffin wax became the major constituent of modern candles.

Whatever the ingredients, the temperature of a candle flame reaches no more than 1400 °C. Simple attempts to increase the size of a candle flame generally lead to less efficient combustion, which implies more soot instead of more light. This is because the rate of burning is determined by the rate at which oxygen can reach the wax vapours and not by the rate of vaporization. An increase in flame temperature and light output can only be achieved through some drastic changes of design. Although candles have been available for more than 2500 years, for most of this time only rich and powerful people had them in sufficient quantity to avoid the need to synchronize their lives with the natural rhythms of light and dark. It was not until the 19th century that artificial sources of light became commonplace for the average person.

In the second half of the 18th century, Antoine Laurent Lavoisier made a number of contributions to science, including the clarification of the chemistry of combustion. Unfortunately he was deemed to belong to an unacceptable social class at the time of the French Revolution, and in 1794 the guillotine detached his head. However, the new understanding of the importance of oxygen in combustion enabled others to design lamps with higher flame temperatures and greater light outputs.

There are a number of ways to make it easier for air to reach the centre of the flame, so that the fuel is burnt more efficiently. One of the earliest was the cylindrical wick for oil lamps, devised by the Swiss Aimé Argand. This played an important part in the development of brighter lights for lighthouses, a topic discussed in chapter 8. The airflow was improved further by surrounding the flame with a glass cylinder, which functioned not only as a chimney but also as a protection against gusts that might extinguish an unshielded flame.

Up to this stage all lamps had incorporated their own fuel supply. Around the beginning of the 19th century it was realized that the by-products of the manufacture of tar by heating coal in the absence of air were valuable fuels. The liquid known as paraffin could be carried to the place of use and delivered to the flame under pressure, whereas coal gas (town gas) could be stored in a central reservoir and distributed by pipes to lamps at a considerable distance from the reservoir. With a gaseous fuel, wicks were not needed. Owners and managers of mines began using coal gas for lighting their own houses and offices before the end of the 18th century. Street lighting using coal gas began to appear in London and Lancashire early in the 19th century. Gas lighting was installed in the House of Commons in 1838.

An important advance in gas burners is associated with the name of Robert Wilhelm Bunsen, a professor at the University of Heidelberg and a chemist of international renown. The technician who actually created the first burners, of a type still found in many school laboratories, rarely gets any credit for his contribution. Actually Bunsen's primary requirement in the 1850s was for a very hot flame with low intrinsic luminosity to enable him to study the colour of light emitted by the vapours of different metal compounds. He discovered two elements in the alkali metal group, caesium and rubidium, through the blue and red colours they imparted to a flame. Bunsen burners introduce air into the flowing gas shortly before the point of combustion, thereby making efficient use of the fuel and achieving a hotter flame. The flame does not emit much light or produce much soot because it contains hardly any unburnt carbon.

For general illumination, white light could be obtained by applying the hot flame to a small piece of some refractory oxide. This discovery is often attributed to the Cornish inventor Sir Goldsworthy Gurney. During the 1860s many theatres were lit by 'limelight', the visible radiation emitted by calcium oxide (quicklime) at temperatures approaching its melting point of 2615 °C. Gaslights became much brighter after the invention of gas mantles by the Austrian Carl Auer von Welsbach around 1885.

8

The mantles were constructed by impregnating a shaped piece of fine cotton cloth with nitrates of thorium and other metals. When heated in a gas flame, the cotton burnt away and the nitrates decomposed to leave a thin and delicate gauze-like structure consisting mainly of thorium dioxide. This material has a remarkably high melting point, well above 3000 °C, and survives for long periods even when white hot unless subject to mechanical shock. To render the mantle strong enough to survive the journey from factory to user it was strengthened with collodion, which rapidly burnt away on first use. This form of lighting for houses, schools and streets was in widespread use by 1895, and survived in some places until the middle of the 20th century. Gas mantles still provide light for camping, though now they are made from alternative materials consisting mainly of cerium dioxide. Although cerium dioxide has a lower melting point than thorium dioxide, it is preferred because thorium is weakly radioactive.

1.3.2   Arc lamps and filament lamps

Sir Humphry Davy demonstrated in 1808 that a very bright light could be obtained from an electric arc across a small gap between two carbon rods; but it was not until the middle of the 19th century that arc lighting came into widespread use in theatres. Though arc lighting was bright it had three major limitations:

(1) The carbon rods burnt away, lasting less than a hundred hours and requiring frequent resetting.
(2) The high intensity of the arc made unintentional glimpses of it unpleasant and even hazardous.
(3) Reducing the current led to an abrupt and total extinction of the arc rather than a gradual decrease in intensity.

In 1879 Thomas Edison in the USA and Joseph Swan in Britain independently invented and demonstrated filament lamp bulbs, which proved to be a much more reliable and controllable form of electric lighting. By 1882 the lamps were being produced in substantial numbers, and lawyers were delighted by the prospect of costly legal disputes about patent rights. However in Britain litigation gave way to a merger forming the Edison and Swan United Electric Light Company in 1883.

9

The filament is a fine wire made of a refractory material and enclosed in an inert atmosphere in a glass bulb to protect it from oxidation. Because it has an appreciable electrical resistance, it becomes white hot when electric current flows through it. Many metals were tried and found wanting. The early commercial lamp bulbs had filaments made of carbon; but carbon has the disadvantage that its resistance decreases as its temperature rises, so that a fixed series resistor had to be included in the bulb cap, wasting energy. In metals the resistance increases as the temperature rises, so no additional resistor is needed. The problem was to discover a metal filament that was both durable and easy to manufacture. The inventor of the gas mantle, von Welsbach, made the first metal filament lamps from osmium, but this metal is rare and expensive. Tungsten is cheaper and has the highest melting point of any metal. By 1911 techniques for making fine wires of this uncooperative metal had become reliable, so that the way was clear for tungsten filament bulbs to become a widely used source of illumination.

It was not until well into the 20th century that electric light became available to substantial numbers of people. As illustrated in figure 1.3, advertisements for upmarket hotels just before the First World War drew attention not only to their location, gastronomic delights, billiard tables, croquet lawns and tennis courts but also to the presence of electric lighting.

Nowadays an ordinary household light bulb is known in the trade as a general lighting service (GLS) lamp. As mentioned earlier, its tungsten filament may reach a temperature of about 2700 °C, which is almost 3000 K or about half the temperature of the surface of the Sun. At this temperature most of the radiation is in the infrared, the wavelength of maximum intensity being about 1000 nm, as shown in figure 1.2. There is hardly any output in the ultraviolet and violet. This can be verified with a pair of sunglasses made of photochromic material, which goes dark when exposed to UV and violet radiation. Although photochromic lenses become dark outdoors in daylight even on a cloudy day, they do not respond to a bright GLS lamp even when it is very close.

The working life of the GLS lamp is limited by the evaporation of tungsten, a process that becomes faster if the applied voltage is increased so that the filament becomes hotter and brighter.

HOTEL GROSVENOR

SWANAGE

THE POSITION OF THIS HOTEL ON THE SOUTH SIDE OF THE BAY IS UNEQUALLED

Enlarged 1905-6 and standing in Beautifully Wooded Grounds of 2 acres, having upwards of 460 feet frontage, sloping down to the water's edge

Winter terms en pension from half a guinea per day

ELECTRIC LIGHT, CUISINE AND WINES EXCELLENT, STABLING AND GARAGE

Telegrams: "GROSVENOR, SWANAGE"   Nat. Tel. 193, Swanage

Figure 1.3. Seaside hotel advertisement from about 1910. The premier hotel in Swanage was keeping abreast of modern technology, with a telephone and facilities for motorists. Electric light was not to be taken for granted.

As the lamp ages its filament becomes thinner, acquires a higher resistance, attains a lower temperature and emits less light. The filament can function in a vacuum, but a bulb normally contains an unreactive gas such as a mixture of argon and nitrogen. The presence of the gas reduces the rate of loss of tungsten from the hot filament, but increases the rate at which heat is transferred from the filament to the glass. GLS bulbs are normally rated by their electrical characteristics, but the light output is not directly proportional to the electrical power consumption. As figure 1.4 shows, the more powerful bulbs not only produce more light but do so more efficiently. For example, the illumination from two

11

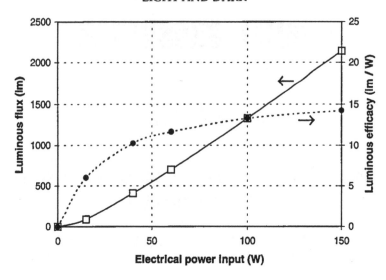

Figure 1.4. Output from ordinary household electric light bulbs. The data points correspond to tungsten filament bulbs widely available in shops and supermarkets. The continuous line and the vertical axis on the left describe the brightness as perceived by an average human eye. The broken line and the vertical axis on the right show how efficiently electrical power is converted into visible light. High power lamps generate more light and are also more efficient.

100 W bulbs is about 1.3 times that from five 40 W bulbs, although the electrical power consumed by the two groups is equal. (This point can be assimilated without understanding the exact meaning of luminous flux and the units in which it is measured. However, it may be useful to point out that this way of describing the brightness of a light emitter involves physiology as well as physics.)

A way of increasing the temperature of the filament without reducing the working life was devised at the laboratories of General Electric in the USA during the 1950s. The tungsten halogen lamp has a much smaller glass bulb, which is filled with the inert gases krypton or xenon (both rarer and therefore more expensive than argon) and a small amount of a halogen (usually iodine, but occasionally bromine). Because the bulb is small, the cost of the gas filling is not excessive and the glass reaches temperatures exceeding 250 °C (523 K). The tungsten filament operates at a temperature around 3300 K, just over ten per cent higher than in a GLS lamp, which results in more intense and whiter light and a higher rate of evaporation of tungsten. Nevertheless, the filament lasts longer because tungsten that condenses on the inside of the hot glass reacts immediately with the iodine to form tungsten iodide vapour. When the tungsten iodide molecules encounter the hot filament they decompose, thereby returning tungsten to the filament. Because the glass becomes so hot, one needs to avoid any contact with skin and other heat-sensitive materials. Glasses with high silica content are required to withstand the temperature and pressure, which explains the alternative name – quartz-iodine lamp. Tungsten halogen lamps containing xenon usually operate at low voltages so that the filament can be very compact without the risk of arcing. They have become standard in modern photographic lighting systems, professional slide projectors and vehicle headlamps.

1.3.3 Gas discharge lamps

During the 20th century a number of alternative types of lighting have been developed. Instead of heating a small amount of a refractory solid to a high temperature, the energy is used in a more selective way, resulting in a much longer working life and considerably higher efficiency. A widely used technique is the passage of electric current through a gas or vapour. An input of 60 W of electrical power into this type of lamp typically produces around ten times as much visible light as the same power supplied to a GLS lamp, or 600 times the light from a single candle. There are two disadvantages: they need additional circuitry to get started; and they emit light at only a small number of discrete wavelengths. Low-pressure sodium-vapour lamps create light very efficiently and so are often used for street lighting. They emit light at only two wavelengths (589.0 and 589.6 nm). As these wavelengths are very close to each other in the yellow region of the spectrum, it is almost impossible to recognize colours. However, the output spectrum is very different for high-pressure sodium lamps. These lamps are less efficient at converting electrical energy into visible light but produce a range of visible wavelengths

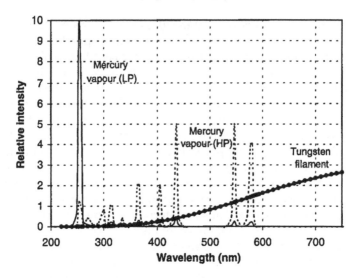

Figure 1.5. Spectra from electrical discharges in mercury vapour. At low pressure the light output is concentrated within a narrow wavelength band in the ultraviolet. At higher pressure the emission shifts to a set of narrow bands within the ultraviolet and visible ranges, shown here by the broken line. In contrast, an incandescent tungsten filament emits a broad range of wavelengths with the maximum intensity in the infrared (beyond the right edge of the diagram).

and so permit moderately accurate colour recognition. Consequently high-pressure sodium-vapour lamps are now often chosen for lighting streets and business premises.

As figure 1.5 shows, pressure affects the emission from mercury vapour lamps too. High-pressure lamps produce visible light directly at five wavelengths (404, 436, 546, 577 and 579 nm) in the violet, green and yellow parts of the spectrum. Fluorescent materials are used to create some red from the unwanted emission in the UV, leading to a better colour balance. The colour rendering and efficiency of high-pressure mercury vapour lamps have led to their widespread use for industrial and street lighting. At low pressure, mercury vapour emits almost exclusively UV radiation, but a number of fluorescent materials are available to absorb the UV and emit light at longer wavelengths in the visible range. This is the basis of the operation of fluorescent tubes. The materials chosen for a particular application depend on the relative importance of accurate colour rendering and of a high overall brightness.

Although a fluorescent lamp may feel cool to the fingers and have gaps in the output spectrum, it can be allocated a nominal colour temperature, which corresponds to the temperature of an incandescent body emitting light subjectively perceived to be similar. Somewhat perversely, the human mind considers reddish light to be warm and bluish light to be cold so that 'warm white' has a colour temperature around 3000 K whereas 'cool white' has a colour temperature around 4000 K.

For the extremely bright white light needed for searchlights, cinema projectors and filming at night, the old carbon arc has been replaced by new forms of electric arc lamp. In xenon short-arc lamps the arc is created in a gap of less than 10 mm inside a silica bulb containing xenon at a pressure of several atmospheres. These lamps can be produced with electrical ratings from a few tens of watts to several kilowatts. They are sometimes used for vehicle headlamps, but generally they are unsuited for domestic use because they have lifetimes of only a few hundred hours, require a very high voltage pulse for starting, contain gas under pressure even when cold and may cause eye damage. The emission is fairly uniform over a range of wavelengths extending from the infrared to the ultraviolet. The output resembles sunlight because it contains substantial amounts of light with short wavelengths.

1.4   Light in art and entertainment

The original purpose of artificial light was to allow productive activities to continue at places and times that would otherwise be simply too dark; but light has also long been linked to leisure and cultural activities. The simple contrast between light and dark fascinated a number of painters including Leonardo da Vinci (1452–1519), Caravaggio (1573–1610), Georges de la Tour (1593–1652) and Rembrandt (1606–1669). There is a style of painting known as *chiaroscuro*, a term derived from the Italian adjectives *chiaro* (light, bright or clear) and *oscuro* (dark, gloomy or obscure). A lamp, candle or occasionally a shaft of sunlight illuminates a small part of an otherwise sombre scene. The Dutch artist Gerrit van Honthorst (1590–1656) became adept at this style when in Rome as a young man. Owing to his dedication to nocturnal scenes he acquired the nickname *Gherardo delle Notti*. He continued to produce this style of picture after his return home. Figure 1.6 (colour plate) shows his 'Christ before the High Priest', in which a single central candle illuminates the scene. The original of this excellent example of *chiaroscuro* art can be seen in the National Gallery in London.

Fireworks produce a fleeting form of artificial light and have an enduring appeal. They originated in China around the 6th century with the discovery of gunpowder, and recipes reached Italy towards the end of the 13th century. The early fireworks were exclusively firecrackers and may have been intended to frighten evil spirits and opposing armies, but their entertainment value was established in Italy by the 15th century. In Europe, Italians continued to be leaders in the development of pyrotechnics. It was an Italian immigrant who founded one of the largest makers of fireworks in the USA, Zambelli Fireworks Internationale based near New Castle, Pennsylvania.

In England the first known record of a formal firework display concerns events at Kenilworth Castle in 1575 for entertaining Queen Elizabeth I. Fireworks remained a highly popular (though expensive and risky) form of entertainment, and skilled pyrotechnicians were lured from country to country to produce increasingly lavish displays. King George II was particularly keen on festivities, and was the British monarch who inaugurated the tradition of royal birthday celebrations twice each year. At his command a famous public display with more than 10 000 fireworks took place in 1749 in Green Park, London to mark the Treaty of Aix-la-Chapelle. It took some six months to arrange, giving time for Handel (Georg Friedrich Händel before he and his name took British nationality) to compose the music that still accompanies firework displays more than 250 years later. The music is now much better known than the political and military events that led to its creation.

Up to the beginning of the 19th century, fireworks were based on chemical compositions that had not changed greatly from the traditional gunpowder mixture of saltpetre (potassium nitrate), charcoal and sulphur. However around 1786 the French chemist

16

Claude Louis Berthollet had discovered how to make potassium chlorate. When this oxidizing agent was used instead of potassium nitrate, flame temperatures increased by about 300 °C. A further increase in brightness was achieved when powdered magnesium and aluminium became available in the 19th century. Another advance in the technology during that century was the addition of various metal halides to pyrotechnic mixtures to produce flames of different colours, for example strontium compounds for red and barium compounds for green. The use of pyrotechnics to create static words and pictures emerged around 1879. The most significant innovation of the 20th century was probably the introduction of computers to control major displays and coordinate the fireworks with music.

In the middle of the 19th century a new and quieter type of light show emerged, involving light guided inside jets of water. Fountains illuminated in this way remained popular features of international exhibitions for much of that century.

Theatres require light to make the actors visible. For thousands of years plays were performed in the open air during the hours of daylight, but outdoors without recourse to any artificial illumination. In contrast, cinematography is totally dependent on the availability of intense artificial light for projection. In addition the recording of the action requires the availability of film sensitive enough for many exposures each second. Consequently this form of entertainment did not arrive until the end of the 19th century, when the first cinema opened in Paris. It was in 1895 that two French brothers patented and demonstrated the cinematograph, which combined the functions of camera and projector. Rather appropriately, their names were Auguste Marie Lumière and Louis Jean Lumière. Early films were taken at 16 frames per second and produced somewhat jerky movements when projected. As film materials became more sensitive, shorter exposures could be used and a recording rate of 24 frames per second became standard. The way in which films are projected has been adapted to the characteristics of the human eye, a topic discussed in more detail in chapter 6.

Light was an essential feature of a form of outdoor nocturnal entertainment first seen in 1952 at the Château de Chambord in France. It was devised by Paul Robert-Houd and is known as *son-et-lumière*. The history of the building is presented with words, music and light without visible performers. This art form retained its French name as it spread across Europe, being seen in Greenwich in 1957 and in Athens in 1959. *Son-et-lumière* later found its way to other continents, being presented at the pyramids at Giza in 1961, in Philadelphia in 1962 and at the Red Fort in Delhi in 1965.

The culmination of the use of artificial light in the sphere of entertainment may be the effects exploited in a discotheque. As this is a topic about which the author is almost totally in the dark, it is time to move on to the next chapter.

2

PATTERNS OF SUNLIGHT

2.1 The year

It might seem a simple matter to define the length of a year as the time taken for the Earth to make exactly one circuit of its orbit around the Sun. The complications commence at the next stage, the definition of the point at which the circuit starts and finishes. There is no absolute frame of reference and all heavenly objects are on the move.

The existence of different definitions of the length of the year was appreciated by the Greek astronomer Hipparchos, who flourished around 135 BC. He lived before accurate clocks were available, so he adopted the solar day as the most reliable time unit and made extensive studies of the lengths of months and years. A comparison of his own astronomical observations with others made a century or two earlier led him to conclude that the positions of stars in the sky at the instant of the autumn equinox were shifting continuously by about one-eightieth of a degree per year. Consequently the duration of a year measured from equinox to equinox was not the same as one measured with respect to star positions. Although his estimate of the shift is now known to have been about 11 per cent too small, his conclusion was a major step forward in astronomy. This phenomenon is now known as *the precession of the equinoxes*, and merits detailed attention.

Stars provide a convenient reference frame for use by astronomers. Weather permitting, they are visible every night and the movements of the Moon and the planets relative to them can be measured. Although the stars themselves are moving at velocities that seem large when given in familiar terrestrial units, they are also a huge distance away, so that for a terrestrial observer the star patterns in the sky alter so slowly that the change is undetectable without the aid of expensive optical instrumentation. If the start and the finish of a circuit are defined by the instants when the angle between the Sun and any chosen star has the same value, the year between those two instants is described as *sidereal*, a term derived from *sidus*, a Latin word for star. Modern astronomers have been able to measure the length of the sidereal year with great accuracy, but four decimal places are sufficient for our purposes. The sidereal year is 365.2564 mean solar days, each mean solar day being exactly 24 hours of Greenwich Mean Time. The word 'mean' in the last sentence is important, as actual solar days vary slightly in duration (a subject discussed further in section 2.3).

However, the rhythms of life on Earth are affected by the seasons and not by the positions of stars in the sky. It is this more appropriate to define a year with respect to the intervals between recurrences of a seasonal event such as an equinox or a solstice. The spring equinox in the northern hemisphere is the usual choice. An equinox is commonly perceived as a time when night and daylight are equally long, but we shall see later that such a definition is unsatisfactory. It is better to regard an equinox as the moment when the Sun is equidistant from both poles and directly overhead at some point on the equator, thereby illuminating the northern and southern hemispheres equally. The interval between successive spring equinoxes is known as a *tropical* or an *equinoctial* year and is equal to 365.2422 mean solar days.

Figure 2.1 illustrates the precession of the equinoxes. The axis of the Earth's rotation is tilted by about 23.5° from the direction at 90° to the ecliptic, the plane containing the Earth's orbit. The magnitude of the tilt hardly changes, but the direction of the tilt rotates very slowly, taking about 25 800 years to complete one cycle. This has two noteworthy effects. The first is that Polaris (the Pole Star) in the constellation Ursa Minor will not always remain directly above the North Pole as a permanent indicator for north in terrestrial navigation. About 12 900 years backward or forward from today, the star Vega was or will be situated over the North Pole. The second is that at successive equinoxes the direction of the Sun is not quite the same every year, so that a tropical year is completed slightly earlier than a sidereal one.

20

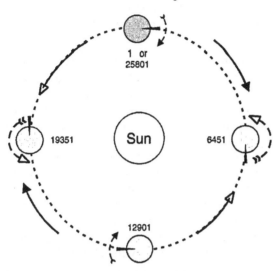

Figure 2.1. Precession of the equinoxes. The Earth, with its axis of rotation projecting from the North Pole, is shown at the northern spring equinox at intervals of 6450 years. The observer is on the north side of the ecliptic, so the South Pole is not visible. The white-headed arrows show the Earth's annual movement and daily rotation, both *anticlockwise*. During a cycle lasting about 25 800 years, the direction of the tilt of the axis of rotation and consequently the position of the Earth at the spring equinox each complete one *clockwise* circuit, as shown by the dark-headed arrows. As a result, the interval between spring equinoxes (the tropical year) is slightly shorter than the time for one circuit relative to star positions (the sidereal year). The diagram is not to scale.

The difference (about 0.0142 days or just over 20 minutes) between a sidereal and a tropical year may seem trivial when considering a single year. Nevertheless, over many years the cumulative effect is important. In a calendar devised to keep in step with sidereal years, the solstices and equinoxes would not remain on fixed dates but would gradually become earlier, shifting by one day in about 70.6 years, one month in about 2150 years, and a full year in about 25 800 years.

The familiar and widely used Gregorian calendar (discussed in more detail in a later chapter) was devised to keep in step with the tropical year and not the sidereal year. With this basis, there is no continuous drift in the dates of equinoxes and solstices, but all the stars occupy slightly different places in the sky at the same moment in successive years, completing one circuit around the sky in 25 800 years.

A quaint result of this shift is the present location of an important reference point in the sky, known as the first point of Aries. The positions of any distant object in the sky can be described in terms of two angles known respectively as the *right ascension* and the *declination*. These two angles are analogous to longitude and latitude in terrestrial navigation. For longitude, the starting line or meridian has been arbitrarily defined as the semicircle that extends from one pole to the other and passes through the old observatory in Greenwich. For right ascension, the analogous reference point in the sky is the position of the centre of the Sun at the spring equinox in the northern hemisphere. This celestial zero point was selected more than two thousand years ago, when it was situated in the part of the sky where the zodiacal sign of Aries began. It is still known as the first point of Aries, even though it has been moved by the precession of the equinoxes into and across the part of the sky allocated to Pisces and is now approaching Aquarius. This shift of star positions poses an awkward problem for astrologers, whose ancient lore and current forecasts are based on interpretations of night skies in an era stretching back to the reign of Nebuchadnezzar in Babylon around 2580 years ago, before Hipparchos made his important discovery.

2.2   Equinoxes and eccentricity

Another major advance in the understanding of the solar system was the discovery that the orbits of the planets around the Sun are elliptical rather than circular. An ellipse can be regarded as a partially squashed circle. The amount of squashing is characterized by a parameter known as the eccentricity, which can have any value between 0 and 1. Two examples are shown in figure 2.2. All ellipses possess two foci located on the long axis on opposite sides of the centre. The eccentricity is the distance between the two foci divided by the length of the major axis, and has a value between 0

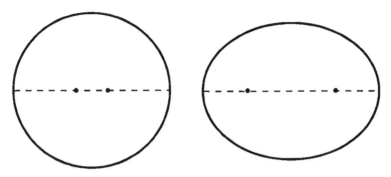

Figure 2.2. Elliptical orbits with different eccentricities. The two ellipses have the position of each focus marked on the major axis. The ellipse on the left resembles the orbit of Mercury and has an eccentricity of 0.2, which means that the distance between the two foci is one fifth of the length of the major axis. The ellipse on the right has an eccentricity of 0.5, which implies that the distance between the foci is half the length of the major axis. The orbits of most planets in the solar system have an eccentricity below 0.1, so that they appear almost circular.

(for a circle) and 1 (for an ellipse so flat as to be indistinguishable from a straight line).

Johannes Kepler was the successor to Tycho Brahe as Imperial Mathematician in the Habsburg Empire, which gave him access to Brahe's extensive and meticulous records of positions of planets in the sky. Kepler was one of Galileo's contemporaries but had the advantage of being based in Prague, sufficiently far from Rome to avoid interference by people determined to uphold Roman Catholic doctrines. In books published in 1609 and 1619, Kepler presented three laws describing planetary motion:

(1) The planets move around the Sun in elliptical orbits with the Sun at one focus.
(2) The orbital velocity decreases as the distance from the Sun increases, so that the radius vector sweeps across equal areas in equal times.
(3) The square of the time required for one circuit is proportional to the cube of the mean radius of the orbit.

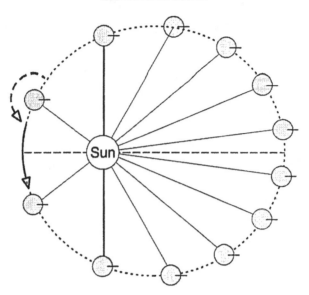

Figure 2.3. Effect of an elliptical orbit on the rate at which the sun's direction changes. The positions of a hypothetical planet with a fixed oblique axis of rotation during one circuit are shown at 12 equal time intervals. The eccentricity of this orbit (0.39) is abnormally high to illustrate the effects more clearly. The Sun lies off-centre at one focus on the long axis of the ellipse. The progress of the planet is not uniform, but is fastest when the planet is closest to the Sun. According to Kepler's second law, the areas of all the drawn segments are equal. The two positions corresponding to the equinoxes are shown by the thick black lines on opposite sides of the Sun. Due to the eccentricity, the time intervals between the equinoxes are far from equal. The two arrows show both the orbital movement and the rotation of the planet on its axis to be anticlockwise.

Figure 2.3 demonstrates the Second Law, by showing the uneven progress of a hypothetical planet in an orbit with an abnormally large eccentricity.

The Third Law implies that an outer planet such as Neptune takes much longer to complete an orbit than an inner planet such as Venus. Not only does Neptune have a greater distance to travel, but it also moves more slowly along its path.

Kepler's Laws are valid for all the planets, including those discovered after his death. Originally the laws had an empirical basis only, but in 1685 Isaac Newton published his theory of gravitational forces, from which Kepler's Laws can be deduced. The size and the eccentricity of the orbits vary from planet to planet. The orbits of Venus and Neptune have eccentricities below 0.01, which means they are almost circular. On the other hand, the orbits of Mercury and Pluto have eccentricities greater than 0.2, which makes them conspicuously elliptical.

The Earth's orbit has an eccentricity of 0.0167. Although this orbit is much less eccentric than those drawn in figures 2.2 and 2.3, the effects of the elliptical shape are not insignificant. The distance from Earth to Sun is at its least (about 147 million kilometres) in early January and at its greatest (about 152 million kilometres) in early July. This means that the intensity of solar radiation, averaged over the whole area of the Earth, is about 7 per cent higher in January than in July.

The eccentricity also influences the times at which events occur. The Earth makes a 360° circuit around the Sun every year, which implies an average change of direction of 0.986° per day. The elliptical nature of the orbit produces a cyclic variation in this change, with a maximum of about 1.019° per day in early January and a minimum of about 0.953° per day in early July. The different rates mean that the Earth's movement to the opposite side of the Sun occupies about 7.5 days less between September and March than between March and September. This inequality is partly compensated for by the structure of the Gregorian calendar, which always has 184 days between 21st March and 21st September, leaving 181.2422 days as the average length of the other six-month period. Because the variable progress around the elliptical orbit has a larger effect than the calendar, the equinox in March normally occurs on a date two or three days earlier in the month than the equinox in September. The ellipticity of the Earth's orbit also influences the length of a day and the time of the true local noon, topics that are considered in section 2.3.

The direction of the long axis of the Earth's elliptical orbit is not fixed relative to a star-based framework, but turns anticlockwise for an observer on the north side of the ecliptic. However, the rotation of the axis is so slow that it takes over 110 000 years to complete one cycle. This rotation is significant for astronomers, but need not be discussed further here.

2.3   The length of a day

The first problem with the word 'day' is its ambiguity. Some languages are inherently confusing because they are the same noun for the period between sunrise and sunset as for the longer period that includes darkness as well as light. This ambiguity is common in west European languages, including French and German as well as English. Other languages have two unmistakably different words, such as *dag* and *dygn* in Swedish. In this book the interval between sunrise and sunset is always described as 'daylight', whereas 'day' is reserved for periods around 24 hours. In the next chapter, a similar principle is applied to the Moon, the period between moonrise and moonset being described as 'moonlight' even when it overlaps the daylight and the Moon's contribution becomes a barely detectable millionth of the total illumination.

The second problem is that the word 'day' has not always represented a period beginning and finishing at midnight. Midnight may be defined as the moment when the Sun is at its greatest depth below the horizon, but direct observation of the Sun's position is obviously impossible. For studying variations in the length of a day it was and is much easier to measure the intervals between visible events. Astronomers also dislike a change of date in the middle of a nocturnal observation, and for a long time they stubbornly maintained that days began and ended at noon. Eventually, they reluctantly fell into step with the majority and agreed that the astronomical day 31st December 1924 would last only 12 hours. In this book, however, we shall consider the length of a solar day as the interval between one noon and the next.

The use of Greenwich Mean Time as a worldwide standard gives the impression that every day lasts exactly 24 hours. In reality the 24-hour period is merely an average value for days defined in one particular way. The fixed day length in the GMT system has been adopted for convenience and the word 'mean' is significant. In this section the Earth is treated as rotating at a constant rate about a fixed axis passing through the north and south poles.

Any quibbling about whether the rate really is constant and the axis fixed will be deferred until section 2.5. In spite of this simplified description of the rotation, the behaviour of the solar system ensures that the time required for one rotation is not an entirely straightforward matter.

The basic problem is associated with the choice of the reference frame required to define the start and finish of one revolution of the Earth on its axis. As with the length of the year, discussed in section 2.1, there are alternative definitions, which produce different results for the length of a day. Stars provide a convenient reference frame. For making accurate measurements of the length of a day, the most appropriate stars are those passing directly overhead for an observer on the Earth's equator. In other words, they are situated on or near the celestial equator, which is a great circle in the sky 90° from the celestial poles. These stars sweep across a wide expanse of sky, whereas Polaris is useless for this purpose as its position in the sky hardly shifts at all in one day. Although the star positions are moving slowly in the 25 800-year cycle described in section 2.1, one day is so small compared with 25 800 years that the adjustment in day length associated with inclusion or exclusion of this cycle in the calculation is less than one hundredth of a second per day. In the present context this adjustment can be regarded as negligible. Any distant star, an expression that excludes our Sun, makes one circuit of the sky in 23.9345 hours, a period known as a *sidereal day*. Its length becomes more comprehensible when described as being almost four minutes less than the familiar mean solar day of 24 hours.

The concept of the sidereal day is rather simple and its duration scarcely varies at all. However, the rhythms of daily life are determined by the movement of the Sun across the sky and not by the positions of distant stars. Consequently the solar day is a more familiar concept, but it is neither as regular nor as straightforward to measure accurately as the sidereal day. The length of a solar day can be directly measured if defined as the time interval between two successive occasions when the observed height of the Sun above the horizon is at a maximum. In astronomical jargon this is known as the *solar transit*, but in more colloquial speech it is the 'local noon'. It does not occur at the same time each day on a clock synchronized with GMT, because an individual solar day can be several seconds longer or shorter than the

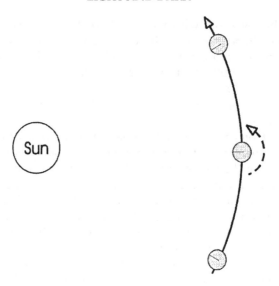

Figure 2.4. Effect of orbital movement on day length. The Earth is shown at intervals of thirty solar days during part of its orbit. The meridian is shown as a line directed towards the Sun, which implies that it is noon at the meridian. The observer is on the north side of the ecliptic (the plane containing the orbit), so that both the orbital motion and the spin appear anticlockwise and the South Pole is not visible. The drawing has a sidereal (star-based) reference frame and is not to scale. A sidereal day corresponds to rotation through 360°. Because the orbital motion changes the direction of the Sun, a solar day requires the earth to rotate through a slightly larger angle, making a solar day almost four minutes longer than a sidereal day.

mean value of 24 hours. The height of the noon Sun also changes day by day, covering an angular range of about 47° between winter and summer, but it is not difficult to cope with this variation.

Figure 2.4 shows why the mean solar day is almost four minutes longer than the sidereal day. The Earth moves in orbit around the Sun as well as rotating about its own axis. The two movements are in the same sense, anticlockwise from a viewpoint above the North Pole. The orbital movement requires the Earth to rotate by about 1° more than a complete circle relative to the stars before the Sun returns to its highest point in the sky.

The fluctuations in the duration of solar days arise from two effects of similar magnitude, though one of them is easier to comprehend than the other.

(1) The orbit of the Earth is elliptical and not circular. In fixed time intervals, the orbital movement sweeps around arcs of constant area but not of constant angle. This produces a cyclic variation in the length of solar days, which are shortest in early July (3.864 minutes longer than the sidereal day) and longest in early January (3.996 minutes longer than the sidereal day). The duration of sidereal days is unaffected by the shape of the orbit as the stars are so far away.

(2) The Earth's equator is tilted at an angle of 23.5° to the ecliptic, the plane in which the orbit lies. This tilt produces seasonal changes in the proportions of darkness and daylight (discussed in section 2.4), but there is also a smaller and less well known effect on the duration of a solar day. A rigorous account of the underlying geometry is outside the scope of this text, but the result is a six-monthly cycle in which the length of solar days is reduced at both the March and September equinoxes and increased at the summer and winter solstices.

The effects of these cycles are shown in figures 2.5 and 2.6. For most people living in the northern hemisphere, it may come as a surprise that the solar day is longest around Christmas. At the end of December, the combined effects of orbital eccentricity and axial tilt extend each solar day by almost 30 seconds relative to the unchanging 24-hour days prescribed by Greenwich Mean Time. Half a minute per day may seem trivial, but the effect of the fluctuating length of the solar day is cumulative. Figure 2.6 shows that on the meridian the Sun reaches its zenith about 16 minutes ahead of 1200 Greenwich Mean Time in early November and about 14 minutes after 1200 GMT in early February. The length of the solar day affects not only the time of the true local noon, but also the sunrise and the sunset, which are subject to other variations discussed in section 2.4.

The implications of the data presented in figure 2.6 may be easier to see in figure 2.7, which shows the varying position of the Sun in the sky at a fixed time of day throughout a year. If the weather were co-operative, you could obtain the curve shown in

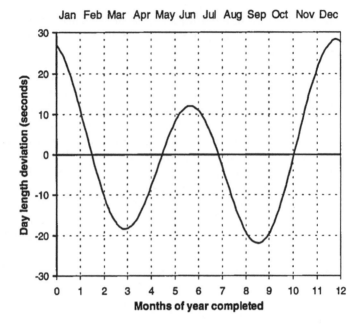

Figure 2.5. Deviation of the length of solar days from the 24-hour average. Noon may be defined as the moment when the Sun is highest in the sky. The time interval between one noon and the next is a solar day, which varies throughout the year. The shortest solar days occur in the middle of September, whereas at the end of December they last almost half a minute longer than the mean of 24 hours.

figure 2.7 by pointing a camera in a south-easterly direction, keeping the film stationary, and opening the shutter for an extremely short exposure at 0900 GMT every day throughout one year. Although the Sun follows the same course across the sky at the two equinoxes, the course is covered about a quarter of an hour earlier on the autumn equinox than on the spring equinox.

By now it should be obvious to you that simple sundials are untrustworthy except on four days a year, though in the three-month period from the beginning of April to the end of June the error is not more than four minutes either way. Of course, it is possible to design a sundial with a more complicated scale taking the astronomical effects into account but there are other pertur-

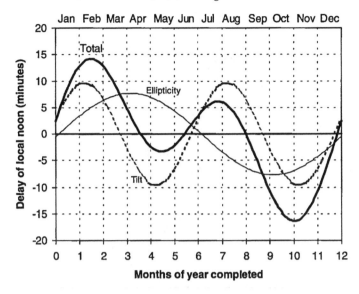

Figure 2.6. Variation of the local noon throughout a year. The time when the Sun is at its zenith each day is affected by the ellipticity of the Earth's orbit and by the tilt of the Earth's rotational axis. The bold curve shows the combined effect. Although the true local noon has an average time of 1200 over the entire year, it occurs around 1214 in the middle of February and 1144 in early November.

bations that need to be considered to achieve a sundial with the highest accuracy.

2.4 The length of daylight

The division of the solar day into a period of daylight and a period of darkness is not constant throughout the year. In the higher latitudes there is a conspicuous and familiar annual cycle, with the hours of darkness longest at the winter solstice and shortest at the summer solstice. The cause of this variation is straightforward and well known. The axis of rotation of the Earth is tilted by almost 23.5° from the direction perpendicular to the ecliptic plane. When the North Pole is nearer to the Sun than the South Pole, the northern hemisphere enjoys longer hours of daylight, and the

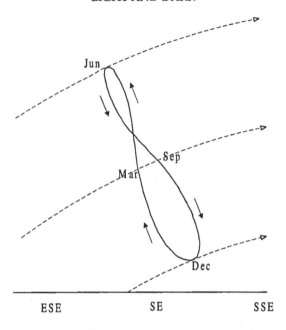

Figure 2.7. Position of the Sun in the sky at 0900 each day of the year. The analemma, resembling a figure of eight, shows the annual variation of the position of the Sun in the sky at a fixed time of day in London. The broken lines indicate parts of the daily circuits of the Sun across the sky at the equinoxes and the solstices. The Sun is high in the sky in June and low in December and the timing of its daily circuit varies. Both this diagram and figure 2.6 show that the time deviations are greatest about seven weeks before and after the winter solstice.

nights are extended in the southern hemisphere. The precession of the direction of the tilt has been described earlier. Although it has a significant effect on the length of a year, the effect on the duration of daylight is negligible as compared with some other phenomena. It is widely but incorrectly believed that at the equinoxes sunset and sunrise are twelve hours apart. Tables listing the times of these events for every day of the year reveal that this is not the case. For example, at the latitude 56° N at the March equinox, sunrise at 0600 is preceded by sunset at 1813 and followed by sunset at 1815. At the same place at the September equinox, sunrise at 0547 is preceded by sunset at 1759 and followed by sunset at

1756. It is apparent that daylight is 10 to 14 minutes longer than twelve hours at this time of year and that night is shorter than twelve hours.

A comparison of daytime length in midsummer with night length in midwinter also reveals a mismatch. At the same latitude, sunrise and sunset are separated by 17 hours 37 minutes of daylight at the summer solstice, whereas sunset and sunrise are separated by 17 hours 3 minutes of night at the winter solstice. During any year there is significantly more daylight than night. There are two factors contributing to this inequality.

(1) The smaller factor arises simply from the definition of the time of sunrise or sunset as the moment when the uppermost part of the Sun first becomes visible over a flat horizon. Sunset is defined similarly as the moment when the Sun finally disappears. Because the diameter of the Sun extends about $0.54°$ across the sky, the centre of the Sun is about $0.27°$ below the horizon at sunrise and sunset.

(2) The larger factor is due to the presence of the atmosphere, which modifies both the timing and the abruptness of the transition from daylight to darkness and *vice versa*. The loss of abruptness is due to scattering, which is discussed later in chapter 5. The alteration of the timing is due to the bending of light, a phenomenon known as refraction. Refraction is due to the lower speed of light in air than in empty space. The ratio of these two speeds is the *refractive index*, which has a value around 1.0003 for air at sea level, though it depends on pressure and humidity. It gradually decreases as the air becomes less dense with increasing altitude. Figure 2.8 shows that when the Sun is close to the horizon, refraction makes it appear higher in the sky than it would be without the atmosphere.

When the two factors are combined, it means that at sunrise or sunset, the true direction of the centre of the Sun is below the horizon at that moment by about $0.82°$, to which refraction contributes $0.55°$ and the Sun's radius $0.27°$. Thus the interval between sunrise and sunset is extended and the interval between sunset and sunrise is shortened. An observer at the equator at either equinox would be able to see the Sun while its true position moved through $181.64°$ of its $360°$ cycle, so that the duration of

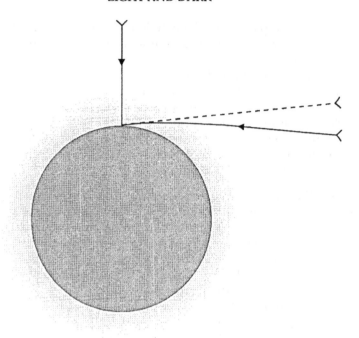

Figure 2.8. Bending of sunlight in the atmosphere. When the Sun is directly overhead, rays of sunlight continue on a straight course through the Earth's atmosphere. When the Sun is near the horizon, the rays are bent continuously as they pass through air of increasing density, so that the Sun appears higher in the sky than its true position. The diagram is not to scale and the bending has been exaggerated for increased clarity.

daylight would be more than 6.5 minutes longer than a half-day and the duration of night 6.5 minutes shorter, producing a mismatch totalling 13 minutes. The extra daylight increases as the observer moves north or south away from the equator, because the Sun no longer moves at 90° to the horizon, but crosses it at an oblique angle. Figure 2.9 illustrates the progress of the setting Sun at 57° N, the latitude of Edinburgh. In this case the extension of daylight is almost twice that at the equator.

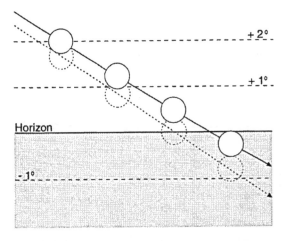

Figure 2.9. Progress of the setting Sun. The circles show positions of the Sun at intervals of 6 minutes around sunset at an equinox at latitude 56° N. As the Sun approaches the horizon, bending of light by the atmosphere increases, making the apparent positions significantly higher than the true positions, shown by the broken circles. Sunset is defined as the instant when the last visible part of the Sun disappears, as shown on the right. In this example, sunset occurs about 6 minutes after the centre of the true Sun has crossed the horizon.

The visibility of the Sun when its true position is slightly below the horizon also extends the area of the Earth from which the midnight Sun is visible. It is not absolutely necessary to be north of the Arctic Circle or south of the Antarctic Circle. At 0.5° (55 km) outside either polar circle, more than half of the disc of the Sun stays above the horizon throughout the night at the summer solstice. At 0.25° outside, the whole disc remains visible. The most northerly parts of the mainland of Iceland extend to within a few kilometres of the Arctic Circle. If the weather is unusually cooperative, you can see the midnight Sun from Raufarhöfn, a little port with a hotel, a population of about 500 and an air link with Akureyri. Crossing the sea to Grimsey, a small island actually on the Arctic Circle, is unnecessary, though taking tourists on trips to this island remains a useful source of income for the Icelanders.

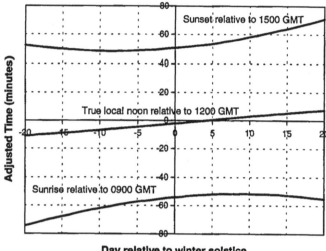

Figure 2.10. Times of sunrise, true local noon and sunset around mid-winter. The variation in the times of three daily events are shown for the period from 1st December to 10th January on the meridian at latitude 52° N. The time at which the Sun reaches its zenith becomes later each day, because each solar day lasts more than 24 hours at this stage of the year (see figure 2.5). At the winter solstice the interval between sunrise and sunset is shortest, but the earliest sunset is 8 or 9 days earlier and the latest sunrise is 8 or 9 days later.

The day with the shortest daylight occurs at the winter solstice, but as figure 2.10 shows, it does not coincide with the latest sunrise or the earliest sunset. The offset exists because the solar days are longest in late December, exceeding the 24-hour mean day length by almost half a minute, as explained earlier and illustrated in figure 2.5. The long day means that true local noon measured in GMT gradually becomes later as December progresses, as shown earlier in figure 2.6. This progressive shift also applies to the sunrise and the sunset, which are both becoming later by almost half a minute per day at the winter solstice. It is the combination of the change in the length of the day and the change in the length of daylight that is responsible for the separation of the dates with the earliest sunset and with the latest sunrise.

The amount of the offset decreases as the latitude increases, but in southern England the earliest sunset is around 12th December and the latest sunrise is around 30th December.

An analogous but smaller offset occurs around the summer solstice. Figure 2.5 shows that the excess length of solar days in June is less than half that in December.

2.5 The length of a second

Homo sapiens must have been aware of years and days as units of time ever since our species came into existence. Shorter and more arbitrary units of time were introduced as civilization advanced. It was the Babylonians, with their counting system of sixes and tens, who first established the 24-hour day, the 60-minute hour and the 60-second minute. In the metric system, the second has always been the fundamental unit of time, but the definition of the second has undergone some considerable changes during the past hundred years. With the developing need for greater accuracy in timing, the limitations of a time scale based on astronomical cycles became clear, so that today the fundamental unit of time is based instead on the physics of caesium atoms.

As we have already seen, the duration of actual solar days has a cyclic variation, with about 50 seconds difference between the longest and the shortest days of any year. This regular cyclic change is only a minor problem, as it is possible to consider all the solar days in a year or in many years and to calculate an average. The time interval corresponding to the mean solar day divided by 86 400 (which is 24 × 60 × 60) was defined as a second. Such a definition was adequate for astronomers in the 19th century and would satisfy most ordinary people today.

The use of the mean solar day explains the presence of the 'mean' in 'Greenwich Mean Time' (GMT). In this system, every day was deemed to have the same duration, during which interval a fictitious 'mean Sun' makes exactly one circuit of the sky. GMT noon is the instant at which the mean Sun is at its maximum height. We have also seen that the real Sun reaches its maximum height at times that vary throughout the year. You can see from figure 2.6 that the discrepancy between the positions of the real

Sun and of the fictitious mean Sun is zero just four times per year, the instants when sundials tell the correct time.

Originally the time in Greenwich was counted from the GMT noon. This meant that 0600 was six o'clock in the evening and 1800 was six o'clock in the morning. However, on 1st January 1925 the time 0000 was re-allocated to midnight instead of noon, creating a 24-hour clock system similar to that in use today. To avoid confusion between the old and new time systems, the International Astronomical Union recommended in 1928 that the new version should be described as Universal Time (UT) and that the term GMT should be used only for times belonging to the pre-1925 system.

Although astronomers obediently adopted the new nomenclature, it has never entered the vocabulary of the average man or woman. Although officially it does not exist today, the term GMT remains in widespread use and has even survived the closure of the Royal Greenwich Observatory in 1998. The BBC's six pips, introduced on 5th February 1924, are still broadcast as a widely recognized time signal, though the final pip was lengthened in 1972 to make the important one more obvious.

Astronomers need to measure times and angles with a precision far beyond that encountered in normal life. Consequently it was astronomers who first became aware of tiny discrepancies in their timings. With the second defined as an exact fraction of a mean solar day, they found irregular variations in the duration of the tropical year. However, they also realized that the experimental data would indicate that there were irregular fluctuations in the duration of the mean solar day if the second were defined as a fraction of a mean tropical year. In order to decide whether the axial rotation or the orbital movement of the Earth would provide the most reliable basis for a time scale, the astronomers measured the cyclic movements of other bodies in the solar system with great care and assessed all the observed irregularities.

It turned out that the apparent discrepancies in the positions of all such bodies were less when the time scale was based on the assumption of a constant tropical year rather than a constant rate of rotation of the Earth. The conclusion was that the rotation of the Earth is not stable enough to be the basis of the most reliable time scale. The rotational fluctuations have both regular and irregular features, due to a range of physical effects. First, some irregular and as yet unpredictable changes are ascribed to movements of the liquid outer core of the Earth. Secondly, an annual cycle with a shallow minimum at some time in the northern summer has been ascribed to weather patterns that lack symmetry because of the different areas and locations of land and sea in the northern and southern hemispheres. In addition there is a steady lengthening of the day by about 1.7 milliseconds per century. This gradual slowing is partly due to tidal friction, which converts rotational kinetic energy into heat. (This abbreviated list is by no means a complete account of this complex topic.)

A new definition of a second was therefore devised in 1956 and adopted in 1958. In Ephemeris Time (ET) the second was defined with impressive precision as the tropical year divided by 31 556 925.9747. For even greater precision, one particular tropical year was cited in quaint and arcane terms involving the zeroth January 1900. Although ET has a sounder basis and so is theoretically superior to UT, in practice it involves inconveniently long and tedious astronomical observations. Within ten years of the introduction of ET, the second had been redefined again in a manner unconnected with astronomy.

The current definition of a second was adopted in 1967 at the Thirteenth General Conference of Weights and Measures. It involves multiplying a very short time by a very large number rather than dividing a long period such as a year or a day by some other number. A second is now 9 192 631 770 times the period of the microwave radiation produced by spin-flip transitions between a particular pair of states of a caesium atom. The very brief time interval can be produced with incredible accuracy, in an instrument known as an atomic clock. The working of this instrument has nothing to do with the movements of the Earth and Sun, and can be explained only in terms of quantum mechanics, a topic outside the scope of this book. A good account can be found in a recent book by Tony Jones entitled *Splitting the Second*.

International Atomic Time (the abbreviation TAI comes from the French *temps atomique international*) is a superbly stable standard. Nevertheless it is not perfectly adapted for everyday use over a long period. Observers equipped with clocks that simply count TAI seconds would see a drift in the times of noon, as defined by either the real Sun or the fictitious mean Sun used in GMT. Expressed in a more dramatic way, at some stage in the distant future Greenwich would be in darkness while such a clock showed noon. To maintain the usual relationship between clock times and daylight hours, clocks that count TAI seconds are adjusted by the introduction of an extra second, roughly eight or nine times per decade at present. The presence of the extra second is marked by a seventh pip in the radio signal. This usually happens at midnight at the end of December or June, but only as required and not following a prescribed and rigid scheme akin to the occurrence of 29th February in the Gregorian calendar. It is predicted that leap seconds will be needed more often in future, possibly reaching fifteen per decade around 2050. With the leap seconds, the system is known as Co-ordinated Universal Time or UTC.

We are thus left with a strange situation. The rotation of the Earth has been found too irregular to be a basis for a reliable and modern definition of a second. Nevertheless the clocks that we actually use are adjusted occasionally to keep them synchronized with the rotation of the Earth. Fewer leap seconds would have been needed if the TAI second had been made 1 part in 40 million longer, so that it resembled a UT second rather than an ET second. The day is still the *éminence grise* of time.

3
MONTHS AND MOONLIGHT

3.1 The lunar month and the lunar orbit

Life on Earth without energy from the Sun is inconceivable, but
life could almost certainly have developed if the Moon had never
existed. The light provided by a full Moon is only about one millionth of the light coming directly from the Sun, but it is more than
ten thousand times greater than the light from Sirius, the brightest
'fixed' star in the sky. Although the Moon is so much less bright
than the Sun, it has influenced animal behaviour and fascinated
humans since time immemorial. The lunar cycles affect not only
the darkness of the night but also the level of the sea, setting the
rhythms of life on the seashore. So this chapter will consider not
only the daily, monthly and longer variations of moonlight but
also some characteristics of tides.

It was mentioned earlier that the orbit of the Earth around
the Sun and the rotation of the Earth on its axis are in the same
direction, namely anticlockwise for an observer above the North
Pole. The Moon, like most of the satellites of planets in the solar
system, moves around its parent planet and rotates on its axis in
the same direction, although the axes of rotation of the two bodies
are not exactly aligned.

The period of the Moon's orbit around the Earth depends on
the way in which the start and finish of a cycle are defined. Two
ways of defining the monthly cycle are shown in figure 3.1. If
stars are used as markers, the time for the Moon to make a circuit is a *sidereal month*, which lasts for 27.3217 mean solar days.
The sidereal month may be considered as the basic cycle, but it is
not the cycle that is most obvious to a terrestrial observer. People

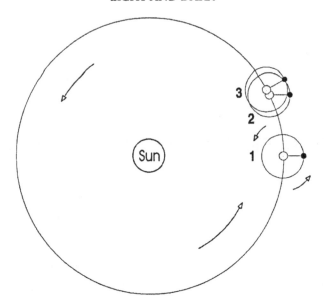

Figure 3.1. Sidereal and synodic lunar months. In positions 1 and 2 the direction of the line from Earth to Moon is the same relative to distant stars or to the edges of the page. The time required to move from 1 to 2 is 27.3217 mean solar days or one sidereal month. In positions 1 and 3, the Sun, Earth and Moon are in line. The time between these two positions is 29.5306 mean solar days or one synodic month. The arrows show the anticlockwise motions of the Earth and the Moon, when viewed from a distant position above the northern hemisphere. The angles are correct, but the distances and sizes are not to scale.

and animals are more aware of the interval separating two successive new Moons, when the Moon's position in the sky is closest to the Sun and the nights are darkest, or of the identical interval between two successive full Moons, when the Sun and Moon are in opposite parts of the sky and moonlight is at its brightest.

The time taken to return to a particular phase of the Moon is 29.5306 mean solar days and is known as a *synodic month* or *lunation*. It is longer than the sidereal month because the Earth and Moon are moving around the Sun, and so the Moon has to travel further than one circuit around the Earth to repeat an alignment with the Earth and the Sun. In sidereal terms, in an average synodic month the Moon travels in its orbit around the Earth through about 389.105°, which corresponds to 1.080 85 revolutions. The difference between sidereal and synodic months resembles on a larger scale the difference between sidereal and solar days. It requires only simple arithmetic to calculate that a year contains 13.368 sidereal months and 12.368 synodic months. It is no accident that these numbers differ by 1. As we shall find in the next chapter, there are various ways in which months and years are adapted to produce calendars in which the number of months per year is an integer.

The orbit of the Moon is often regarded as a circle centred on the Earth, but this is an inaccurate description. Whereas the Earth has hardly any effect on the position of the Sun, which is about 330 000 times more massive, the Earth is only about 80 times more massive than the Moon and is appreciably affected by it. The centre of gravity of the Earth–Moon combination is displaced from the centre of the Earth by about 4700 km, equivalent to about 73 per cent of the Earth's radius. Figure 3.2 illustrates the movements of the Earth and the Moon around this point during one sidereal month. Both paths are elliptical, although the deviation from a perfect circle is barely perceptible in figure 3.2. Although the eccentricity is only 0.055, it is more than three times as great as the eccentricity of the Earth's orbit around the Sun. This implies that the distance between Earth and Moon is not constant but can be up to 5.5 per cent more or less than the mean value. The ellipticity is also associated with varying speed around the orbit, so that the Moon positions shown at equal time intervals are more widely spaced at the bottom of figure 3.2 than at the top.

3.2 The lunar nodes and their rotation

The concept of the lunar nodes and their rotation through a complete circle every 18.60 years is a matter of which most people are unaware. Nevertheless the cycle can be detected and its length measured without sophisticated instruments. It is therefore plausible that its existence was known early in the development of astronomy, although the reason for it was not fully understood until much later. The rotation of the lunar nodes provides the key

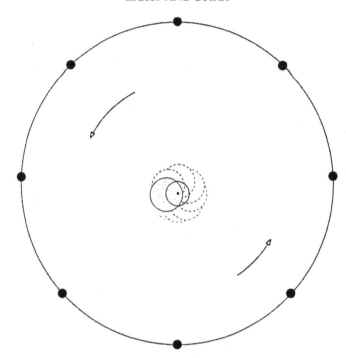

Figure 3.2. Movements of Earth and Moon around their common centre of gravity. The common centre of gravity is shown by the small central dot, which is at a shared focus of both elliptical orbits and not at the centre. Positions of the Earth and the Moon relative to it are shown at eight equal time intervals within one sidereal month. The grey circle shows the position of the Earth when the Moon is on the right. The progress of the Moon is slightly more rapid at the bottom of the diagram than at the top. The diameters of the Moon, the Earth and the track of the Earth's centre are correctly scaled, but the lunar orbit has been reduced in size to fit on the page.

to understanding the height of the Moon relative to the horizon and the timing of eclipses.

The cycle becomes apparent from any simple study of the positions of moonrise or moonset at a particular time of year over a period of many years. For example, observations of the direction of the full Moon as it approaches moonset around midwin-

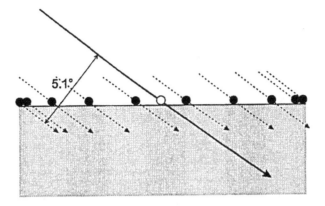

Figure 3.3. Variations of the path of the midwinter full Moon around moonset. The horizon is shown for an observer in Britain looking approximately northwest. The path of the setting Sun at midsummer, shown by the bold line, remains the same. In contrast, the path of the setting full Moon around midwinter changes from year to year. Ten paths, separated by nine one-year intervals, are shown. The average time for a complete cycle from one extreme to the other and back is 18.6 years.

ter would reveal a sequence of changes from year to year similar to that shown in figure 3.3. Whereas the directions of the midsummer sunrise and sunset are maintained, there are regular variations in the directions of the rising and the setting full Moon around midwinter. The cycle time of 18.60 years is independent of the observer's location but the distance along the horizon between the two extremes depends on latitude.

It has been suggested that there are features of Stonehenge connected with observations of this cycle. Some holes near the northeast entrance once contained the bases of wooden posts that could have been markers for recording the directions of midwinter moonrise. At the centre of Stonehenge, the horseshoe pattern of bluestones, brought from southwest Wales around 2500 BC, contains 19 stones. Nobody knows why this number was chosen, but it could have allowed these stones to be used for counting 19 years as an approximate cycle of the lunar nodes. Another speculative suggestion concerns the large circle of 56 Aubrey holes lying outside the sarsen ring but inside the circular ditch

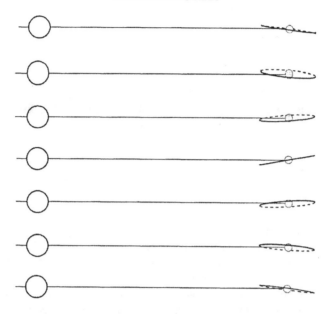

Figure 3.4. The change of direction of tilt of the Moon's orbit. The Earth is shown on the right, moving away from the reader at the same time of year at three-year intervals. The plane of the lunar orbit is never parallel to the ecliptic, which is at 90° to the page, intersecting at the black line. The angle of tilt is constant but the direction of tilt rotates, almost completing one cycle in the eighteen years illustrated here. The diagram is not to scale and the angle of tilt has been exaggerated for clarity.

and banks. Neither the function of the holes nor the reason for their number has been established, but the astronomer Sir Fred Hoyle pointed out in the 1960s that a marker that was moved three holes per year would complete a circuit in 18.67 years, which is a fairly good match to the cycle of 18.60 years.

The changing direction of the moonrise and moonset occurs because the lunar orbit does not lie in the same plane as the ecliptic, which contains the path of the Earth around the Sun, but is at an angle of about 5.13° to it. Although this angle remains almost constant, the direction of the tilt rotates continuously at 19.35 degrees per year. Figure 3.4 shows the direction of tilt at intervals of

46

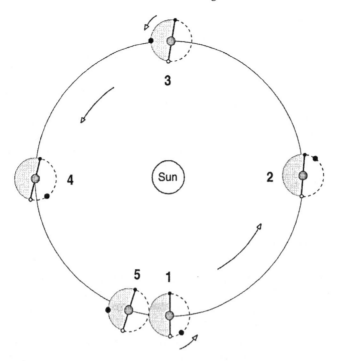

Figure 3.5. Rotation of the lunar nodes and an eclipse year. The positions of the Earth numbered 1 to 4 correspond to intervals of one quarter year. The orbits of the Moon at each position of the Earth are shown. The lunar orbits are anticlockwise and do not lie in the ecliptic, but cross it at places known as nodes, shown by the smallest black and white circles. The straight lines show the alignments of the two nodes, which rotate clockwise at 19.4° per year. The alignment towards the Sun shown in position 1 recurs in position 5 after an interval of 346.62 days, which is known as an eclipse year. The angles in the figure are correctly scaled but the distances are not.

exactly three years during a period of eighteen years, slightly less than the time required to complete one full cycle.

Figure 3.5 illustrates the same phenomenon viewed from a different direction and at intervals of less than a year. During one quarter of their annual circuit around the Sun the Moon and Earth make 3.342 circuits around their common centre of gravity, which is why the Moon is shown shifted by about 123° per quarter. The Moon crosses repeatedly from one side of the ecliptic to the other and back again. In figure 3.5 the grey and the white semicircles show on which side of the ecliptic each part of the lunar orbit lies. The time between two successive crossings of the same type is 27.212 days. This is slightly less than a sidereal month because the direction of tilt is rotating slowly in the opposite sense to the orbiting Moon. The places where the path of the Moon crosses the ecliptic are known as *lunar nodes*. The straight black lines in figure 3.5 show the alignment of the lunar nodes rotating clockwise at 19.35 degrees per year. This rotation is a key feature in the timing of eclipses.

3.3   The lunar day

It was pointed out in the previous chapter that the duration of a solar day could be measured most readily as the time interval between two successive occasions when the Sun attains its highest position in the sky. Similarly, the duration of a lunar day for a terrestrial observer can be measured as the time between two successive lunar transits, the moments when the Moon is at its highest position in the sky. The mean length of such a lunar day is 24.8412 hours, which is just over 50 minutes longer than the mean solar day. One of the most obvious manifestations of the lunar day is the time interval between high tides. These normally occur twice per lunar day, implying an average interval around 12 hours 25 minutes. We shall return to the tides later in section 3.6.

The difference in length between lunar and solar days means that a terrestrial observer sees the Moon moving westward across the sky at a rate that is on average about 3.4 per cent slower than that of the Sun. This implies that the Moon becomes farther east in the sky at a given time on successive days. The slower movement of the Moon across the sky also means that in each synodic month the number of moonrises is one less than the number of sunrises.

We noted earlier that fluctuations of no more than half a minute in the duration of solar days were due to the elliptical orbit of the Earth and the tilt of the plane of the equator relative to the ecliptic. The ellipticity and the tilt of the lunar orbit affect the duration of the lunar day in a similar manner. Figure 3.6 shows

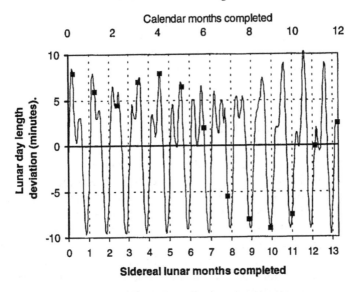

Figure 3.6. Deviation of length of a lunar day from the average during one year. The lunar day is the time interval between moments when the Moon is at its highest in the sky. The mean is 24.8412 hours, but fluctuations up to ten minutes either side of the mean occur during each sidereal lunar month. Both the magnitude and the frequency of the fluctuations are much greater than those for the solar day, shown in figure 2.5. Black squares indicate when the Moon is full. These events are separated by one synodic lunar month, which is longer than one sidereal lunar month.

the variations in lunar day length during the year 1993. There are conspicuous differences between this diagram and the variability of solar days, shown in figure 2.5. First, a lunar day can deviate by as much as 10 minutes from the average, a deviation about 20 times larger than that for a solar day. Second, there are many more maxima and minima in figure 3.6 than in figure 2.5. The basic pattern in figure 3.6 is repeated every sidereal lunar month, but the form of the monthly cycle is not maintained exactly from month to month, nor is it repeated exactly in successive years. Two factors are responsible for the greater variability in the duration of lunar days. First, the orbit of the Moon is more eccentric than the Earth's orbit around the Sun. Second, the effects are more concentrated because the number of lunar days per month is less than the number of solar days per year. The lunar day length cycle matches the sidereal lunar month and is not synchronized with the phases of the Moon, which are associated with the synodic lunar month.

3.4   The length of moonlight

The term 'moonlight' describes the time when any part of the Moon is above the horizon. This definition is independent of the phase of the Moon and applies even for a new Moon that is barely visible because its dark side is towards the Earth and its apparent position in the sky is close to the Sun. The causes of prolonged visibility are the same as for the Sun and have already been described. The bending of light by atmospheric refraction contributes about 0.55° when the Moon is close to the horizon, and the angular diameter of the Moon means that the upper edge is visible if the centre is less than 0.26° below the horizon. Other phenomena, however, have more effect on the duration of moonlight than on the duration of daylight.

The first of these is that the closeness of the Moon (as with any other satellite) makes it visible for a shorter time. Figure 3.7 shows how the proximity of an artificial geostationary satellite in a circular orbit of radius 42 155 km substantially reduces the fraction of the Earth's surface from where it is visible. Neglecting the refractive effect, the geostationary satellite is invisible from about 55 per cent of the equator (197° out of 360°). The distance of the Moon (on average 384 400 km) is about nine times that of a geostationary satellite and 60 times greater than the radius of the Earth. At the equator, the Moon is concealed for 181.9°, which corresponds to an extra 0.95° at both moonrise and moonset. This can be translated as a reduction of 7.6 minutes in each period of moonlight. At higher latitudes the reduction is greater, exceeding 12 minutes in Britain. The reduced time of visibility due to the proximity of the Moon outweighs the sum of the extensions already noted. The overall result is that the average duration of moonlight is somewhat less than half the average lunar day, whereas the average duration of daylight is more than half the solar day.

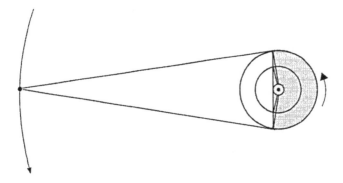

Figure 3.7. Effect of proximity on satellite visibility. The artificial satellite on the left is in a geostationary orbit, meaning that it is in the equatorial plane at 42 155 km from the centre of the Earth, and has the same angular velocity as the Earth's rotation. This satellite can be seen from less than half of the earth's surface. At the equator it is visible from less than 163° of the total circumference, the angle decreasing as the latitude increases. It is not visible from any area close to the North Pole, shown here as a black dot. The Moon is about nine times further away, so the area from which it can be seen is larger but still less than half of that of the Earth's surface.

The familiar annual daylight cycle, in which the Sun is above the horizon for longer in summer than in winter, has a lunar analogue. When the Moon is closer to the North Pole than to the South Pole, the northern hemisphere has longer periods of moonlight than average and the southern hemisphere has shorter periods. The moonlight cycle, shown in figure 3.8, occurs monthly and not annually, so the changes in the duration of moonlight from day to day are much larger than the changes in the duration of daylight. The difference between the maximum and the minimum durations of moonlight depend on latitude, as with daylight. Midway between the two extremes of the monthly cycle when the moonlight lasts for just over 12 hours, the change is rapid; in Britain the duration of moonlight may increase or decrease by as much as one hour per day. This may be contrasted with the maximum rate of change in the amount of daylight in spring and autumn, which is only about 4 minutes per day.

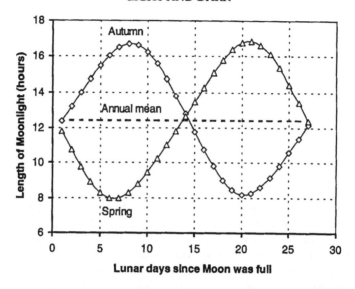

Figure 3.8. Variation in length of moonlight each month. The time between moonrise and moonset varies in a manner resembling the seasonal variations in daylight, but with two major differences. The moonlight cycle is complete in one sidereal lunar month instead of in one year. The average duration of moonlight is appreciably longer than 12 hours.

As mentioned earlier, the apparent movement of the Moon across the sky is slower than that of the Sun and the lunar day has a mean duration of 24.8412 hours, just over 50 minutes longer than the mean solar day. The lengths of individual lunar days are variable and can be almost 10 minutes longer or shorter than the mean, owing to the eccentricity and tilt of the lunar orbit. Consequently the lunar transit on successive days, measured on a GMT clock, may occur from 40 to 60 minutes later. The shift to later times on successive days also applies to the times of moonrise and moonset. Figure 3.9 shows how the time of moonrise becomes later as a result of the combination of the varying lunar day and the varying length of moonlight. At 52° N moonrise is less than 20 minutes later each day in one part of the month, but about two weeks afterwards it is almost 90 minutes later each day. The cycles match the sidereal lunar month and not the phases of the Moon.

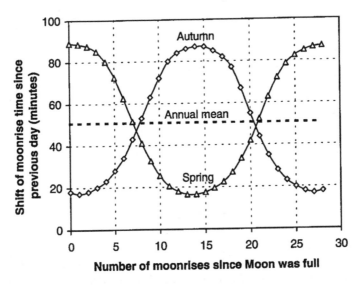

Figure 3.9. Daily shift of time of moonrise at 52° N. On average the Moon rises just over 50 minutes later each day, but there are wide variations of the time-shift within each month. The left and right extremities of the diagram correspond to the full moon, and the centre to it to the new moon. Because the number of sidereal lunar months in each year is one more than the number of synodic lunar months, the moonrise time-shift cycle does not keep in step with the phases of the Moon. As a result the monthly cycle in the autumn is the inverse of that in the spring.

At higher latitudes, the difference between the extremes is greater than that shown in figure 3.9; near the Arctic and Antarctic Circles one can observe moonrise at the same time on successive days.

It has always been convenient for agriculture that the slow changes in the time of moonrise coincide with the full Moon in autumn. The phenomenon became known as the 'Harvest Moon' and was a benefit for the human race for thousands of years before the invention of electric light and mechanical harvesters. Crops had to be harvested around the autumn equinox, but the hours of daylight were often inadequate for completing the task. Work could continue into the evening for the few days around the full Moon in autumn because the Moon rose as the Sun set and the

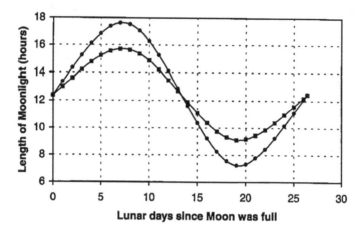

Figure 3.10. Short and long term variations in the length of moonlight. The two curves have the same form as the autumn curve in figure 3.8 but show the minimum and maximum monthly variations in the autumn at latitude 52° N. The cycle lasts 18.6 years and is due to the rotation of the direction of the tilt of the lunar orbit.

time of moonrise was changing by less than 20 minutes each day. This slow change arises from a combination of three features, namely autumn, full Moon and moonrise. If any one of these is replaced by its opposite (spring, new Moon or moonset), the rate of change becomes rapid. Two substitutions of opposites, of course, result in another slow rate of change. The effect is similar in the southern hemisphere, although here autumn occurs around March.

The length of moonlight shown in figure 3.8 is for a particular year, but a long-term study of the maximum and minimum amounts of moonlight in each monthly cycle reveals the existence of another periodic variation. The monthly range is not constant, but oscillates between maxima and minima, illustrated in figure 3.10. This effect is due to the rotation of the lunar nodes, described in section 3.2, and is related to the changing direction of the rising and setting Moon.

3.5 Eclipses and Saros cycles

3.5.1 Eclipses and history

It would have been straightforward for prehistoric people to become aware of the daily, monthly and annual rhythms of sunlight and moonlight. On the other hand, they would probably have been surprised and frightened by unexpected disappearances of the Sun. Anyone able to predict these events successfully would have seemed both awesome and wise. However, the accumulation of enough data for demonstrating any kind of pattern would have been a formidable task. Ancient observers and data collectors had several handicaps besides the presence of clouds. Partial solar eclipses could pass unnoticed because the Sun was too bright to look at directly. Total solar eclipses would be impossible to ignore, but these are limited to narrow strips across the Earth's surface, and at any one location it is rare to have more than one such eclipse in a lifetime. In ancient times both human mobility and communications were poor. Collecting information would have been much easier for lunar eclipses than for solar eclipses. Although lunar eclipses are less common and less spectacular, they last longer because the Earth's shadow is bigger than the Moon's shadow. Furthermore, because of the Earth's rotation, each lunar eclipse is potentially observable from more than half the area of the Earth's surface during some of the period of totality. Some aspects of solar and lunar eclipses are illustrated in figures 3.11 and 3.12.

The Moon does not become completely invisible as it passes through the Earth's shadow, but darkens to a coppery red because some sunlight still manages to reach it, owing to scattering by the Earth's atmosphere. An observer on the Moon at that time would experience a total solar eclipse, but would see the dark side of the Earth surrounded by a reddish ring.

3.5.2 The Saros cycle

There are conflicting opinions about when and where the periodic recurrence known as the Saros (usually pronounced like 'say Ross') cycle was first recognized. Cuneiform writing on clay tablets suggests that it may have been the Chaldaeans, who lived

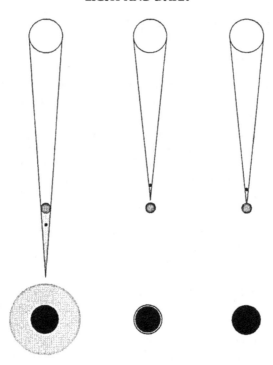

Figure 3.11. Lunar and solar eclipses. The lower part of the diagram shows a terrestrial observer's view of an eclipse, due to the shadows shown above. A total lunar eclipse (left) lasts much longer than a solar eclipse because the Earth's shadow is much wider than the Moon. The umbra (total shadow) of the Moon does not reach the Earth's surface every time that the Sun, Moon and Earth are aligned. Annular solar eclipses (centre) are more common than total solar eclipses (right). Even when the Moon is close enough to hide the Sun completely, the total solar eclipse is limited to a narrow strip of the Earth's surface. The diagram is not to scale.

around 300 BC in what is now southern Iraq. They noted that one lunar eclipse was often followed by another 6585 days later. (We shall see later why it was easier to spot the repetition of lunar eclipses.) More than two thousand years later Edmund Halley produced the first really coherent account of the Saros cycle. He became a professor at the University of Oxford in 1704 and As-

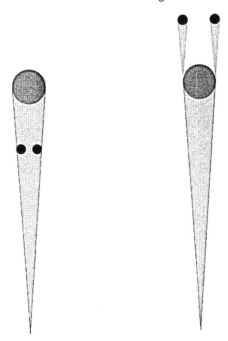

Figure 3.12. Positions of Moon, Earth and their shadows in lunar and solar eclipses. The movement of the Moon is in a direction at 90° to the plane of the page. The range of positions of the Moon for a lunar eclipse (left) is narrower than the range of positions for a solar eclipse (right). Weather permitting, a lunar eclipse is visible from almost half of the Earth's surface, but a solar eclipse is visible from only a small area. These features mean that lunar eclipses are less frequent, but immobile people can experience them more often. The diagram is not to scale.

tronomer Royal in 1720. He was particularly interested in eclipses because he had the rare good fortune to be active during a period when southern England enjoyed two total solar eclipses (from different Saros cycles) separated by only nine years. The length of the Saros cycle is now known to be 6585.32 mean solar days, a little shorter than the time required for one revolution of the lunar nodes. The period may also be expressed as 18.031 tropical years. In calendar years, the duration can be one day shorter or longer than the normal 18 years 11.32 days, depending on

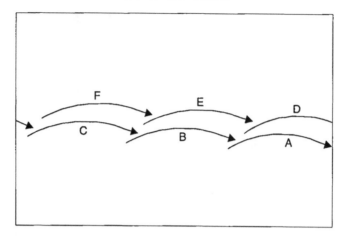

Figure 3.13. Eclipse tracks within part of one Saros sequence. The rectangle represents a map of the Earth with tracks of six successive solar eclipses in the middle of one Saros sequence. The arrows show that the shadow of the Moon travels eastward during each eclipse, but each eclipse track in the sequence lies about 115° to the west of its predecessor. In an even Saros sequence like this there is also a small shift northward. In an odd sequence the westward shift is accompanied by a small shift southward. Eclipse F would occur 90.155 tropical years after eclipse A.

whether the 29th February occurs five or four times during the cycle and whether or not the observer crosses the International Date Line while taking the shortest route to see the next eclipse in the sequence. An account of the International Date Line can be found in the final section of chapter 4.

The eclipses forming any one Saros sequence occur regularly, but the places where the successive solar eclipses of one sequence are visible move around the globe in a corkscrew manner, as illustrated in figure 3.13. Each successive eclipse path in the same sequence lies about 115° further west. This westward shift is a direct consequence of the fraction of a day included in the 6585.32 days of the Saros cycle, and the change of longitude is simply 360° multiplied by 0.32. In addition to the substantial westward movement there is also a small movement northwards for even sequences or southwards for odd ones. The northward or southward drift implies that each sequence has a limited life, being born at one pole and dying at the other.

The total solar eclipse visible in Cornwall on 11th August 1999 continued across Europe and part of Asia. It was preceded by a total eclipse visible in eastern Siberia on 31st July 1981 and will be followed by a total eclipse visible across the USA on 21st August 2017. These three eclipses are only a small part of the Saros sequence numbered 145, extending from January 1639 to April 3009. About forty-two Saros sequences of solar eclipses exist concurrently, producing around 238 eclipses per century, though only a minority are total. The co-existence of so many sequences means that some places may enjoy two total eclipses in less than a decade. One such fortunate location lies a little to the north of Lobito on the coast of Angola, with total eclipses on 21st June 2001 and on 4th December 2002, a mere eighteen synodic months later. Another lies in the middle of Turkey about 250 km east of Ankara, with total eclipses on 11th August 1999 and on 29th March 2006.

Nowadays it is fairly easy to grasp that solar eclipses must be associated with new Moons. By definition, successive new Moons are one synodic month apart. As explained in section 3.1 and illustrated in figure 3.14, a terrestrial observer sees the Sun moving westward across the sky each day at a slightly faster rate than the Moon. The new Moon occurs as the Sun overtakes the Moon, but the apparent paths across the sky usually do not coincide at the moment of overtaking. A solar eclipse occurs only on the rare occasions when the Sun and Moon pass simultaneously through a point in the sky where the two paths cross. Expressed in a different way, the alignment required for a solar eclipse occurs only when the Moon is new as it makes its transit through a lunar node.

In figure 3.5 the Earth positions 1 and 5 both have the black lunar node lying on the line connecting the centres of the Earth and the Sun. The return to this alignment takes less than a tropical year, because the lunar node is rotating slowly in the opposite (clockwise) sense to the (anticlockwise) movement of the Earth. The mean interval between the alignments involving the same node is 346.62 days, a period known as an eclipse year. On average during this period the direction of the line from Earth to Sun changes by 341.64° and the direction of the line connecting the two nodes changes by 18.36°. The total of these two angles is

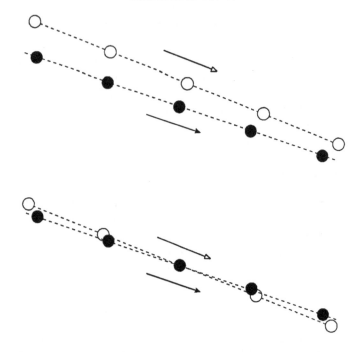

Figure 3.14. Tracks of the Sun and Moon across the sky. Both diagrams show five moments in the progress of the Sun (open circle) and the new Moon (dark circle), as seen by a terrestrial observer. In the upper diagram the overtaking does not occur at the point where the two tracks in the sky cross. The lower diagram shows one of the rare occasions when the overtaking coincides with one of the two crossing points, resulting in a solar eclipse.

360°, so the realignment is achieved. If both nodes are considered, the interval between alignments is 173.31 days.

As solar eclipses can only occur at the time of a new Moon, the interval between them has to be a whole number of synodic lunar months. The Saros cycle of 6585.32 days is 223 synodic lunar months. This happens to be almost the same as 19 eclipse years, which amount to 6585.78 days. The difference of 0.46 day is the cause of the small change in latitude at which the repeated eclipses can be seen. Depending on which of the two nodes is involved, the latitude at which the repetition of an eclipse is observable moves either always northwards or always southwards. The journey lasts between 1244 and 1514 calendar years, which implies between 69 and 84 Saros cycles. At the start of a sequence all that occurs is a partial eclipse near one pole and at the finish all that occurs is a partial eclipse near the other pole. Total and annular eclipses occur in the middle of a sequence.

3.5.3 Total and annular solar eclipses

Even when the alignment is precise, the eclipse may not be total. The Moon's umbra, the small grey shadow in figure 3.11, is on average 374 300 km long. The average separation of the centres of the Moon and Earth is 384 400 km. Although an observer on the surface of the Earth where the Moon is directly overhead (and therefore closest) may be almost 6400 km nearer to the Moon than an observer with the Moon near the horizon, the adjusted distance is greater than the average length of the umbra. In these circumstances perfect alignment produces an annular eclipse with the outer edge of the Sun still visible. Total eclipses are only possible because the orbits are elliptical so that the distances of the Moon and the Sun fluctuate. As mentioned in section 3.1, the lunar orbit is more than three times as eccentric as the orbit of the Earth around the Sun and the distance from Earth to Moon can be up to 5.5 per cent greater or smaller than the average. Consequently, the variations in the separation of Moon and Earth (up to 21 100 km either side of the average) are more than three times as great as the variations in the length of the Moon's umbra (up to 6300 km either side of average). Expressed in terms of the angular diameter seen by a terrestrial observer, the average Sun (0.53°) appears wider than the average Moon (0.52°). However, the fluctuations in the angular diameter due to the elliptical orbits are approximately ±0.01° for the Sun and ±0.03° for the Moon. As a result, the relative widths are sometimes reversed, enabling the Moon to obscure the Sun totally. The longest total eclipses of the Sun occur when the Moon is at its closest and the Sun at its farthest, making the Moon appear to be about 6 per cent wider than the Sun. As mentioned earlier, the distance between Earth and Sun is greatest early in July. Consequently, long total eclipses are most common during July and the adjacent months. The Saros sequence numbered 136 currently produces unusually long eclipses, including the eclipse seen in Africa for over seven minutes on 30th June 1973, followed by eclipses lasting almost seven minutes on 11th July 1991, visible in Mexico, and on 22nd July 2009, visible on the west side of the Pacific Ocean. Both these intervals are 18 calendar years plus 11.32 days, but the second one includes a westward shift across the International Date Line that is compensated by the fifth leap year in the interval between 1991 and 2009.

The distance from Earth to Moon can now be measured with extraordinary accuracy with the aid of lasers and the reflectors left on the surface of the Moon by the Apollo astronauts. Such measurements indicate that the distance is increasing by around 38 mm per year. This implies that total eclipses are very slowly becoming shorter and less common, while the proportion of annular eclipses is increasing. If this continues for another 620 million years, the Moon will be so far away that it will never be able to obscure the Sun completely and there will be no more total solar eclipses.

It is worth noting that within solar Saros sequences a total eclipse is usually followed by another total eclipse, whereas an annular eclipse is usually followed by another annular eclipse. This feature must have made the existence of Saros sequences more obvious, but it is due to a fortunate coincidence. The distance between the Earth and Moon varies cyclically with a period known as an anomalistic month, which is 27.554 55 mean solar days. By chance, 239 anomalistic months add up to 6585.54 mean solar days, which is a remarkably close match to the 19 eclipse years (6585.78 days) or the 223 synodic months (6585.32 days) that determine the Saros cycle. Consequently, the Moon returns to almost the same distance from the Earth as the eclipses of a Saros cycle recur.

A total solar eclipse may be experienced for longer if the observer moves with the Moon's shadow as it speeds eastward at about 3400 km/h. If the Sun and Moon are overhead, the Earth's spin moves the surface in the right direction, with the speed varying from about 1670 km/h at the Equator to zero at the poles. It has been calculated that the maximum possible duration for an observer on the Equator at a fixed longitude would be about 7.5 minutes. The least additional eastward speed needed to keep up with an eclipse is about 1730 km/h, which is within the capability of some aircraft. On the occasion of the solar eclipse visible in Africa in June 1973, a Concorde aircraft remained in the area of totality for more than 70 minutes, about ten times longer than the period of totality experienced by a static observer in the optimum location. Unfortunately for the passengers the Sun was high in the sky and this aircraft has no roof windows, making observation of the solar corona extremely difficult. The same problem arose during another Concorde flight prolonging the eclipse in August 1999.

The Sun and Moon are usually seen moving from east to west across the sky and are below the horizon when moving from west to east. As mentioned earlier, the Sun crosses the sky slightly more quickly than the Moon, so a lunar shadow usually tracks across the Earth's surface in a roughly easterly direction. The eclipse paths shown in figure 3.13 are all consistent with this rule, but exceptions are possible. During summer far from the equator, the Sun remains above the horizon for much longer than twelve hours, so that some or all of its eastward movement is revealed. If a solar eclipse occurs when the local time is closer to midnight than to midday, the lunar shadow moves westward. An example is the annular eclipse that will occur in the early morning of 31st May 2003. It will begin in northern Scotland and proceed northwest to Iceland and then Greenland. Around Inverness the maximum coverage of the Sun will almost coincide with sunrise and the eclipsed Sun will be within a few degrees of the horizon for all observers. Cricket enthusiasts may regard such a rare event as an astronomical googly.

3.6 Tides

In general terms, it is well known that the changes in sea level known as tides are related to gravitational effects of the Moon and the Sun, with the Moon having the larger effect. It may therefore appear paradoxical that the Earth is subject to a larger gravitational attraction toward the Sun than toward the Moon. Another initially puzzling aspect of tides is that the water level is high not only in the part of the sea closest to the Moon but also on the opposite side, so that usually there are two high and two low tides per lunar day.

The strength of the gravitational fields due to the Sun and the Moon can be calculated using a simple formula devised by Isaac Newton, involving the distances and masses of the Sun and Moon. Although the average distance of the Sun is about 390 times the average distance of the Moon, the Sun has a mass about 27 million times that of the Moon. Consequently the Sun exerts a gravitational pull on the Earth that is almost 180 times as great as the pull exerted by the Moon. (This ratio is simply $27 \times 10^6 \div 390^2$.) It is important to understand that a spatially uniform gravitational field would produce no tides, because all parts of the planet would be affected equally. The distortion of the surface of the sea results from two concurrent effects. First, the near side of the Earth experiences a stronger gravitational field than the far side does. A calculation based on Newton's formula shows that the difference across the diameter of the Earth is more than twice as great for the Moon because it is so much closer than the Sun. Secondly, the Earth and Moon are both moving around their common centre of gravity, as illustrated in figure 3.2. This motion produces a centrifugal effect dependent on the distance from the centre of the rotary movement. Sea farthest from the Moon experiences both the greatest centrifugal effect and the weakest pull towards the Moon. Consequently the sea level is high here as well.

The difference between sea levels at high and low tides depends upon the directions in which the Moon and Sun lie. Spring tides occur when the effects of the Sun and the Moon are additive, producing large daily changes in sea level. The Sun, Moon and Earth are almost aligned at new moon and full moon, so that the spring tides happen twice within each synodic lunar month. When the Moon appears as a semicircle, the directions of the Sun and Moon are at 90° and the tidal effect of the Sun opposes that of the Moon. This situation produces the neap tides, during which the difference between high and low water is at a minimum. The connection between spring tides and the season known as spring is more apparent in the English language than at the seashore. The confusion between spring tide and the season of spring does not arise in French (*marée de vive eau* and *printemps*), in German (*Springflut* and *Frühling*) or in many other European languages.

The alignment of Sun, Moon and Earth has a cumulative and not an immediate effect on the height of tides, so that the greatest difference between high and low water does not occur exactly at new Moon or full Moon, but two or three days later. This delay in reaching the greatest amplitude of the daily tidal oscillation has some similarity to the timing of the maxima and minima in the daily or the annual temperature cycles. The hottest part of the day usually occurs in the early afternoon and the hottest part of the year is normally about a month after the summer solstice.

The variations in the length of the lunar day, shown in figure 3.6, mean that tides have an inherent irregularity, which is increased by variations of weather and the shape of the coastline. The shape and location of the coastline can drastically modify both the amplitude and the form of the tidal cycle. In the Mediterranean Sea the difference between high and low tide is normally less than half a metre, even for spring tides. In contrast, the Bay of Fundy, lying between the Canadian provinces of Nova Scotia and New Brunswick, has spectacular tides, with the sea level sometimes changing by more than 15 metres. Almost as impressive are the spring tides around St Malo in Brittany, where the sea level can change by 12 metres.

In many places around the Gulf of Mexico only one high and one low tide occur per lunar day at some parts of the monthly cycle. On the other hand there are places where high water may occur four times per lunar day. A cycle in which each high tide is split into two is common on the coast of southeast Dorset and is illustrated in figure 3.15. The data for Dover in this figure illustrate that low water does not necessarily occur exactly midway between two high waters.

Even dry land is not immune from tidal effects, though the distortions are on a much smaller scale. Even so the radius of the Earth at one land location on the equator changes by about 20 cm.

When devotees of astrology are asked to explain how planetary positions might influence events on Earth, they sometimes invoke a comparison with tides. Only an elementary knowledge of the theory of gravity is needed to evaluate the magnitude of the tidal effects on Earth due to the planets of the solar system. The planet that is able to provide greatest possible tidal effect is Venus, our nearest planetary neighbour. It gets as close as 41 million kilometres from Earth about every 584 days. Although Jupiter has a mass 390 times greater than that of Venus, it is always more than 600 million kilometres away and therefore produces a smaller

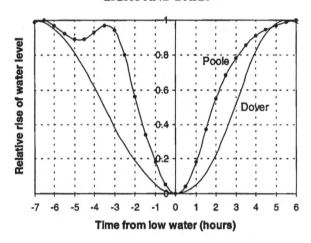

Figure 3.15. Water levels in one cycle midway between spring and neap tides. Real tidal cycles rarely have simple and symmetrical forms and they may have different forms for spring and neap tides. On the coast of southeast Kent near Dover the tide usually rises more rapidly than it ebbs. On the coast of southeast Dorset near Poole there is often a double high tide. An average tidal cycle lasts about 12 hours 25 minutes.

tidal effect. Even at the closest approach of Venus, the resulting gravitational field gradient at the Earth is less than one ten thousandth of that due to the Moon. It is also less than that due to a ten-ton object five kilometres away. This suggests that the gravitational effects arising from movements of trains and aircraft in our vicinity are more influential than those associated with the movements of planets. Predictions based on airline and railway timetables could become an inspiring feature of tabloid newspapers in the new millennium.

4

HISTORY, DATES AND TIMES

4.1 Solar calendars

The Gregorian calendar that is in widespread use today has been devised to match the tropical solar year to a high degree of accuracy. It aims at preventing long-term drift in the dates of the summer and winter solstices and the spring and autumn equinoxes. It can be classified as a purely solar calendar because it makes no attempt to keep in step with the cycles of the Moon. Because it is linked to a particular part of the lunar cycle, the date of Easter changes from year to year. The history of the development of the Gregorian calendar provides insight into some of its quirky features, such as the derivation of the name of the twelfth month from the Latin word for ten. It also explains why the October Revolution occurred in November 1917.

By the year now described as 46 BC, the Roman calendar was in chaos. Years had contained only 355 days, a fairly good match to twelve synodic lunar months. To compensate for the shortness of the basic year, leap years containing a short extra month were required, but the extra month had been applied irregularly and insufficiently. Lack of understanding of the solar cycles was partly responsible, but the occasional adjustments of the calendar were sometimes arranged for political purposes, such as changing the duration of someone's term of office. Julius Caesar, then holding the office of *Pontifex Maximus*, decided that remedial action was essential and sought the advice of Sosigenes, an astronomer in Alexandria. Julius Caesar's interest in this city was not entirely due to its eminence in mathematics and astronomy. Alexandria was where he had started his notorious affair with Cleopatra, the young queen of Egypt who was not totally enthusiastic about marriage to her younger brother.

In order to restore the start of the following year to the expected season, the year 46 BC was extended to 445 days, thereby earning the title of the Year of Confusion. The lengths of subsequent years were to be reckoned according to a new system, which became known as the Julian calendar. Each year contained twelve months with durations that we would recognize today, although the days were numbered by a peculiar Roman system that required counting backwards from three reference points in each month. The system was cumbersome because the reference points moved; for example, the Ides fell on the 15th day of March, May, July and October but the 13th day of other months. Once in four years, the sixth day before the Calends (first day) of March was immediately repeated. Relics of this double sixth remain today in various languages. The ordinary words for a leap year are *une année bissextile* in French and *uno anno bisestile* in Italian. The adjective *bissextile* can be found in English dictionaries, though nowadays it is encountered more often in crosswords than in conversation.

Neither Julius Caesar nor the corrected calendar survived unscathed for long. He was murdered in Rome on the Ides of March in 44 BC and afterwards the senators demonstrated that a successful career in politics was compatible with incompetence in science and mathematics. For the following 36 years, one year in three was deemed to be a leap year with 366 days. Augustus subsequently reduced the accumulated error by omitting some leap years and from 8 AD the Julian calendar was functioning more or less as had been intended.

The additional day in February once in four years made the duration of an average Julian year 365.25 days. This is just over 11 minutes longer than 365.2422 days, the tropical or equinoctial year discussed in chapter 2. Consequently the equinoxes and solstices shifted gradually to earlier dates within the sequence of Julian years, by about one day every 128 years. The resultant advance of the spring equinox to 21st March was considered at the Council of Nicea in 325 AD, but the outcome was a simple acceptance of the situation and there was no adjustment of the frequency of leap years. With only minor changes to simplify the way in which the days in each month were labelled, the Julian calendar remained in use up to the 16th century in most of Europe, and survived even longer in countries where Roman Catholicism was not the dominant religion and papal decrees were regarded with disdain.

By the middle of the 13th century the Julian calendar had accumulated a conspicuous error, a matter that disturbed a scholarly English Franciscan friar. Roger Bacon was interested in numerous topics now classed as science, including the anatomy of the eye, the origin of rainbows and the properties of gunpowder. Although his fellow friars actively hindered his activities in such fields, Pope Clement IV became aware of his remarkable talents and invited him to communicate his views on numerous topics, including the flaws in the calendar. Unfortunately Clement died shortly after Bacon's lengthy manuscripts arrived in Rome and they were used to discredit Bacon as a heretic. The problem due to the excess of leap years in the Julian calendar was allowed to grow for another three centuries before anyone in the Vatican became seriously concerned. In the 1570s Pope Gregory XIII appointed a commission to consider the question and later consulted a Neapolitan astronomer, Luigi Lilio Ghiraldi, about ways to restore the equinoxes to somewhere near their dates at the time of the Council of Nicea. The outcome of the discussions was a bold and drastic change. Pope Gregory decreed in 1582 that ten adjacent dates should disappear, so that the day after Thursday 4th October became Friday 15th. To maintain the equinoxes at the required dates, he announced shortly afterwards that there should be no leap year at the end of each century unless the year number could be divided exactly by 400. On this basis there are 485 leap years in 2000 years, making the average interval between leap years 4.1237 years. This implies that the average year length is 365.2425 days, a very good (but not quite perfect) match to the length of the tropical year. The discrepancy between the tropical year and the average year in the Gregorian calendar is only 0.0003 days per year, so that more than three thousand years will elapse before the cumulative error reaches one day.

The Gregorian calendar was swiftly adopted by countries with rulers subscribing to the Roman Catholic faith. Italy, France, Spain and Portugal all changed to the new calendar in 1582, followed by several other countries in central Europe during the next five years. However Luther, Calvin, Knox and King Henry VIII had challenged the authority of the Pope earlier in the 16th century for a variety of reasons, and the movement known as the Reformation had altered the religious landscape of northern Europe. Countries in which Protestant churches were dominant demonstrated their independence from the Vatican by retaining the Julian calendar for many more years. In Switzerland the adoption of the Gregorian calendar was a localized, gradual and prolonged process, canton by canton. Great Britain and its overseas territories did not adopt the Gregorian calendar until 1752. By that time the required adjustment had grown to eleven days, and so 2nd September was followed immediately by the 14th, provoking protests from people who believed that their lives had been shortened by the manoeuvre. The authorities took a more realistic view and required convicted prisoners to serve the full number of days implied by their original sentence. Sweden adopted the new system in 1753 and the Lutheran states in what later became Germany in 1775.

The adoption of the Gregorian calendar spread eastwards in a sporadic and irregular way. Japan adopted it in 1873 and Egypt in 1875, ahead of a number of countries in eastern Europe that were influenced by the doctrines of the Orthodox Church. In Russia the Julian calendar outlasted the Tsars, with the result that the October Revolution (when the Bolsheviks seized power) appears in history books today as an event in November 1917. Greece caught up with the rest of Europe in 1923. Many Eastern Orthodox churches have adopted a partial reform, celebrating Christmas on the Gregorian 25th December but still using the Julian calendar to set the date of their movable religious festivals. Later in this chapter we shall consider the rules that link lunar cycles with the timing of Easter.

4.2 The Roman Catholic Church and the development of astronomy

In the 16th and 17th centuries the Roman Catholic Church contrived to be simultaneously both innovative and conservative in its attitude to astronomy. It led Europe towards the adoption of an accurate calendar, and was not opposed to all attempts to provide accurate data as a basis for astronomy.

Cathedrals and churches were substantial and stable structures, making them potentially useful frameworks for astronomical measurements. In 1576 a simple instrument was installed in the cathedral in Bologna under the direction of a Dominican mathematician named Ignazio Danti. It was effectively a large pinhole camera, with a small hole in the roof producing an image of the Sun on the floor, in which a long metal strip was inserted. The strip, known as a meridian, began directly below the hole and extended northwards like a long ruler, with distances marked on it. As the image of the Sun moved eastwards across the strip around noon each day, the crossing point and the size of the image were noted. The size of the image was greatest in midwinter when the Sun was at its lowest and the image fell at the greatest distance from the beginning of the strip. The daily positions of the Sun's image, recorded over a number of years, could have provided an improved estimate of the length of the tropical year, information needed to devise the Gregorian calendar before its introduction in 1582.

In the middle of the following century the cathedral was extended and Gian Domenico Cassini, a professor of mathematics at the University of Bologna, was given the task of improving and using the instrument. Its size can be judged from the horizontal graduated scale extending for some 67 metres, and from the 260 mm diameter of the image of the midsummer Sun. Cassini subsequently concentrated his attention on astronomy. In 1669 he was enticed to Paris, where he acquired French nationality and altered his name to Jean Dominique Cassini. He became director of the Paris Observatory, a post held subsequently by his son, his grandson and his great grandson.

The observations in Bologna were continued after Cassini had left. In the following century the data were used to demonstrate that the tilt of the Earth's axis of rotation was decreasing, although at an extremely slow rate. The rate of decrease is now known to be 0.013° per century, smaller than the original estimate, but the ability to detect a change in this parameter by the cathedral's instrument was an impressive achievement.

In 1543 the Polish astronomer Copernicus had published a book promoting the idea that the planets moved in circular orbits around the Sun. The concept of the Sun and not the Earth being at the centre was deemed contrary to the scriptures and therefore a heresy. Even Protestants, including Martin Luther, promoted the geocentric dogma as enthusiastically as the Pope. A papal decree in 1616 explicitly prohibited the publication of works advocating the Copernican view. Astronomers needed the skill of expressing their conclusions ambiguously and avoiding direct confrontation. Unfortunately Galileo Galilei lived rather too close to Rome, and expressed his views in a manner so forthright that it was imposs-ible for the Roman Catholic Church to ignore them.

Galileo had been born in Pisa in 1564. His early achievements included correct descriptions of the motion of pendulums and projectiles. In 1592 he was appointed Professor of Mathematics at the University of Padua, and in 1610 he was seconded to the court of the Duke of Tuscany in Florence, where he was able to devote himself to research without the distraction of teaching du-ties. In 1609 he started to develop telescopes that enabled him to observe some of the satellites of Jupiter and the phases of Venus. The varying appearance of Venus could be explained with elegant simplicity by assuming that the planet moved around the Sun in an orbit smaller than that of the Earth. Although he initially sub-mitted to the papal decree of 1616, he published in 1632 a book advocating the Copernican view of the Solar System, and this was seen as a public challenge to the authority of the Roman Catholic Church. He was called before the Inquisition, and under threat of torture retracted his assertions in 1633. He remained under what would now be called house arrest until his death in 1642.

The ban on his book remained in force until 1835. As recently as 1981 Pope John Paul II set up a commission to clarify the events concerning Galileo and the Roman Catholic Church. The validity of Galileo's case was admitted more than a decade later. Some 350 years after his death, the light of reason replaced the darkness of dogma.

4.3 The start of the year

In the early days of the Roman Republic, well before the Christian era, dates in March were regarded as the start of the year. This is why in many European languages the ninth, tenth, eleventh and twelfth months have names related to the Latin words for seven, eight, nine and ten. Although the Romans had moved the start of the year to a date much closer to the midwinter festivities before the Julian calendar was introduced, the concept of the year beginning in spring survived in various forms for another eighteen centuries in Europe.

The choice of 25th March as the date on which the year number increased by one had a basis that was both theological and physiological. This date coincided with the Feast of the Annunciation, which celebrates the start of the life of Jesus *in utero*, a neat nine months before the celebration of his birth on 25th December. Until the middle of the 18th century, England remained ambivalent about the date on which the New Year began, although Scotland had adopted 1st January about a hundred and fifty years earlier, just before James VI of Scotland became James I of England. Consequently there were two versions of the year in simultaneous use in the United Kingdom until 1751. Whereas the historical and Scottish year began at the start of January, the civil and legal year began on 25th March and for some obscure reason the fiscal year started one day later. In the twelve weeks after 1st January, many documents required dates that included both versions of the year number. A possible relic of the springtime start of the year is still evident in the United Kingdom today, with the tax year starting on 6th April, an apparently strange choice. The explanation is that in 1753 the Gregorian 6th April would have been 26th March in the recently abandoned Julian calendar. By postponing the start of the new fiscal year from the Gregorian 26th March until the Gregorian 6th April 1753, the number of days in the fiscal year 1752 was left unchanged in spite of the loss of eleven days in the historical year 1752. This ensured that people with rent or tax to pay could not complain that they were being charged at the full annual rate for a shortened year.

4.4 Lunar and other calendars

The Julian and Gregorian calendars were devised to keep in step with the solar cycles, but the length of a calendar month (about 30.44 days on average) is not closely matched to the interval between new moons, known as the synodic lunar month (29.5306 days). Most calendar systems devised in countries east of the Mediterranean Sea pay more attention to the Moon and use different starting dates for counting the years. In the Jewish and the Muslim calendars, the months contain 29 and 30 solar days alternately. In the Muslim calendar a year contains 12 months based purely on lunar observations. No attempt is made to match the year to solar events, but extra days are incorporated to maintain an accurate match to the synodic lunar month. In 11 years within a 30-year sequence, the addition of one extra day to the final month increases the average length of a month from 29.50 days to 29.5306 days, which is an excellent match to the synodic lunar month. The average Muslim year is therefore 354.367 days, almost 11 days less than the Gregorian year or the tropical year. This means that 34 Muslim years are approximately equal to 33 Gregorian years. Consequently the Muslim festivals are not fixed at any particular season and occur at progressively earlier dates in the Gregorian calendar system.

The Jewish calendar contrives to maintain an average month that matches the phases of the Moon while also achieving an average year that matches the tropical year. This is not a straightforward task, because the time taken by the Earth to complete a circuit around the Sun has no simple relationship with the time taken by the Moon to complete a circuit around the Earth. In fact one tropical year of 365.2422 days contains 13.368 sidereal lunar months or 12.368 synodic lunar months. (The difference of exactly one between these last two numbers is not accidental, but is analogous to the difference of one day experienced by Phileas Fogg in his fictional trip around the world in eighty days.)

Although it is awkward that the number of lunar months of either type in one tropical year is not a whole number, it happens fortuitously and conveniently that there is a way in which the relationship between months and years can be described in terms of integers that are not too large. As long ago as 432 BC an Athenian named Metos discovered that 19 years and 235 cycles of the Moon were virtually identical. Consequently the period of 19 years is now known as a Metonic cycle. With the benefit of accurate modern measurements, 19 tropical years can be calculated to contain 6939.6 mean solar days. This is an extremely close match to the 6939.7 mean solar days that are contained in 235 synodic months or 254 sidereal months.

Because Gregorian calendar years do not allow fractions of a day and contain either 365 or 366 days, 19 years on that calendar can never match 19 tropical years exactly. Most periods of 19 calendar years contain five leap years and 6940 days, but some contain 4 leap years and 6939 days. At the turn of a century, 19 years could include only 3 leap years and hence only 6938 days. Nevertheless, the phases of the Moon on any day of a year normally repeat themselves 19 years later with any errors cancelling each other rather than accumulating over a series of 19-year cycles. The Golden Number of a year indicates the position of that year within the Metonic cycle. Its name derives from the gold inscriptions on monuments in Athens, listing all the dates of full moons over a 19-year cycle.

The average length of a Jewish month is matched to the synodic lunar month by alternating 30-day and 29-day months. The average length of a Jewish year is increased to match the tropical year of 365.2422 days through the insertion of an extra 30-day month in 7 years within each 19-year Metonic cycle. 30 days multiplied by 7 and divided by 19 produces the required average annual extension of just over 11 days. Adjustments of one day are also required, not only for tuning the average month to a better match with the synodic lunar month, but also to avoid starting any year on a Sunday, Wednesday or Friday. This means that a Jewish common year with 12 months can last for 353, 354 or 355 days, and a Jewish leap year with 13 months for 383, 384 or 385 days. The Jewish New Year occurs between 5th September and 5th October of the Gregorian calendar. The concept of a leap month is thought to have been introduced by the Babylonians before 2000 BC, but its use remained haphazard and subject to local variations for a long time.

The calendar used in China was lunar until 1911, when the Gregorian calendar was introduced. Nevertheless the old Chinese calendar is still used for deciding the dates of some social and religious events in various parts of southeast Asia, such as the New Year festivities, which can be at any time between 21st January and 19th February.

There is one major group of events that do not have fixed dates in the Gregorian calendar but depend on the phases of the Moon. Between Shrove Tuesday and Corpus Christi there is a sequence of Christian festivals on dates defined by a particular interval before or after Easter Sunday. As well as Ash Wednesday, Palm Sunday, Maundy Thursday, Good Friday, Ascension Day,

Whit Sunday and Trinity Sunday there are secular holidays that depend on the date of Easter Sunday, which is defined as the Sunday following the first full Moon occurring on or after the 21st March. The first full Moon can occur from 0 to 28 days after the 21st March and the following Sunday can be from 1 to 7 days later. Consequently Easter Sunday can be on any of 35 dates from 22nd March to 25th April. Not all dates within this range are equally probable. Only one combination of delays (0 and 1) can produce the earliest date, and Easter Sunday has occurred on this date only once in the last 200 years. Similarly, only one combination (28 and 7) can produce the last date, and this has happened only twice in the same period. In contrast, any date in the range from 28th March to 19th April can be produced from seven alternative combinations, thereby increasing the probability of Easter occurring on any of these dates. All 23 of them have happened between five and eight times in the last 200 years. Where the Eastern Orthodox churches hold sway, the Julian calendar is still used to decide the date of Easter. In some years this leads to the same result as the Gregorian calendar, but in other years the Eastern Orthodox Easter is one, four or five weeks later. The maximum separation occurred in 2002, when the two Easters were on 31st March and 5th May.

Lunar cycles featured in some Inca records but they played no part in calendars devised by many of the early civilizations in America. The Mayans based their arithmetic on the number 20, which featured in three different calendar systems. For civil affairs they used a calendar with eighteen periods of 20 days plus an extra five days to produce a year lasting 365 days. For reasons that are still obscure, the mystical rituals of the Mayans were based on a 260-day period in which cycles of twenty days and thirteen days were counted simultaneously. This involved changing both identifiers every day and not waiting for one to complete its cycle before altering the other. (We too have independent cycles, with seven names of days that repeat regularly without regard to the day number, the month or the year.) The third system was used for very long times, the largest unit being a Great Cycle lasting $20^3 \times 18 \times 13$ days. This comes to 1 872 000 days, equivalent to a little more than 5125 tropical years. Followers of the Mayans today believe that the completion of a Great Cycle on the Gregorian date 23rd December 2012 will bring cataclysmic changes.

4.5   Time zones

At any given instant, the time of day is different in Tokyo, London and Los Angeles because these cities lie at different longitudes. As the day of the week and the date change at the local midnight, for most of the time the date in Tokyo is ahead of the date in Los Angeles. In practice the boundaries of many time zones are determined by natural and political boundaries as much as by longitude, so that the shapes of the time zones have some strange and intriguing quirks. The International Date Line is perhaps the quirkiest of all.

In a world where politics, communications and transportation were unimportant, the difference between the local time and a standard time such as Greenwich Mean Time could be simply related to the longitude. The change in clock setting would be continuous at a uniform rate of one hour earlier for every 15° farther west and one hour later for every 15° farther east. Travelling westwards at the latitude of London, a degree of longitude corresponds to 70 kilometres, the local time becoming four minutes earlier. The inconvenience of making frequent small adjustments to clocks during a journey eastwards or westwards became intolerable as soon as railways had begun to offer faster journeys, and for more people. It was the need for a national railway timetable that forced the British Government to standardize the time throughout the country.

The idea of broad time zones with clearly defined boundaries is attributed to Sandford Fleming. He was born in Scotland in 1827 and emigrated to Canada in 1845. He served as chief engineer of Canadian railroad companies, including Canadian Pacific, from which he retired in 1880 to become Chancellor of Queens University in Kingston, Ontario. His work had made him aware of the problems of allowing each community to set up its own independent time base, so he proposed a system of zones. Within each zone all clocks would show identical time, which would be exactly one hour earlier or later than the clocks in the adjacent zones to the east or west. At the International Meridian Conference in Washington DC in 1884 the delegates of the 27 nations approved and adopted his proposal. Fleming was granted a knighthood in 1897 and died in 1915.

In practice the time zones are not always 15° wide, nor are they aligned precisely in a north–south direction. They have been adjusted to allow most countries to maintain a single time system within their national boundaries. China is the largest country with a single zone, although there is a difference of about 60° of longitude between the eastern and western borders. Clocks in China are set eight hours ahead of Greenwich Mean Time, an arrangement appropriate for the most densely populated areas located around 120° east. With a similar range of longitude, the main part of the United States (i.e. without Alaska and Hawaii) is divided into four zones, known as Eastern, Central, Mountain and Pacific Time.

The country with the largest number of time zones is Russia. As figure 4.1 shows, Russia has eleven zones with boundaries that are based on political areas and show little connection with the lines of longitude. The most westerly zone is very small and consists of the area around Kaliningrad, which is cut off from the rest of Russia by Poland and Lithuania. Most of the European part of Russia, lying west of the Urals, is in the large second zone. At the western edge of this zone lies the city of St Petersburg, renowned for its 'white nights' in midsummer. The late onset of darkness is partly due to its northerly location and to the clocks being on summer time. However, there is a third factor: St Petersburg is only 30° east of the Greenwich meridian, but its clock setting is more appropriate for a location farther east. Three time zones are very small and restricted to the south of the country. Consequently a journey across northern Siberia requires two-hour clock changes at three of the zone boundaries.

Most time zones throughout the world conform in having the minute hands of their clocks in identical positions, although the hour hands are different. There are a few areas where the official time differs from Greenwich Mean Time by an odd number of half hours. In the western hemisphere this is so for Newfoundland and Surinam. In the eastern hemisphere it is so for Iran, Afghanistan, India, Sri Lanka, Myanmar (formerly Burma), various small islands and parts of Australia.

The mainland of Australia extends across about 40° of longitude, so the presence of three time zones is not surprising. The clocks in the eastern and the western zones conform to Fleming's system and would always differ by two hours if the various states

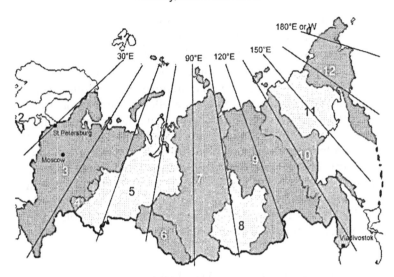

Figure 4.1. Time zones in Russia. The number in each zone shows how many hours ahead of GMT the clocks are during the winter. The boundaries between zones are political and do not follow lines of longitude. In the extreme west, the small area around Kaliningrad has clocks only 2 hours ahead of GMT. The zones 4, 6 and 8 hours ahead are present only in the south. In contrast the zones 10 and 12 hours ahead extend a long way in the north–south direction. The eastern extremity is in the western hemisphere, but the International Date Line has been diverted so that all Russian territory lies to the west of it.

involved could agree about the seasonal clock changes for daylight saving. On the other hand, the central zone does not conform. In South Australia, the Northern Territory and a small area at the western edge of New South Wales the clocks are set in winter to be one and a half hours ahead of those in the west and half an hour behind those in the east. This is not quite as weird as it appears at first, because most of the population of this central strip live in or around Adelaide, which is much nearer to the eastern zone than the western zone. Lord Howe Island belongs to Australia although it is situated in the Tasman Sea about 600 kilometres or 6° from the east coast. It has its own non-conforming time zone, half an hour ahead of the eastern zone.

Nepal displays a rare idiosyncrasy in having the minute hands at 90° to those in other countries. In that country the clocks run five and three-quarter hours ahead of Greenwich Mean Time and fifteen minutes ahead of India, which lies to the south.

At the North and South Poles all lines of longitude meet, the daily course of the sun is almost parallel to the horizon and the concepts of north, south, east and west are meaningless. All time zones become equally valid or invalid, so the clock setting is chosen for administrative convenience. The Amundsen–Scott base at the South Pole is named after Norwegian and British explorers, but is run by the USA. The clocks are in fact set to the same time as in New Zealand, because most of the supplies come from there via the McMurdo base, located between the South Pole and the southwest corner of New Zealand.

4.6 The International Date Line

In a simple and straightforward world, the International Date Line would be expected to coincide with a longitude of 180°, dividing into two the time zone that differs from Greenwich Mean Time by twelve hours. The position of the International Date Line, however, was not defined or agreed at the 1884 conference in Washington, which means that it can be modified by decisions of governments of countries situated close to longitude 180°. The position of the International Date Line can be seen in figure 4.2.

Tonga and some other groups of islands in the Pacific located around 175° west preferred to have the same dates and days of the week as New Zealand although Auckland and Wellington are situated around 175° east. To achieve the desired matching of dates and days, the International Date Line was shifted 7.5° eastwards in the part of the Pacific Ocean lying between 15° and 45° south. This shift had no visible effect on the Tongan clocks, but they are now regarded as being thirteen hours ahead of Greenwich Mean Time instead of eleven hours behind.

In the northern hemisphere the International Date Line performs a zigzag. A westward diversion reaching 170° east at 53° north allows all the Aleutian Islands to have the same days and dates as the mainland of Alaska and the rest of the United States.

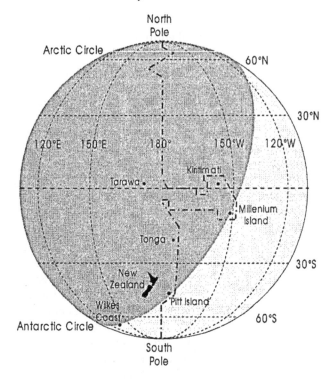

Figure 4.2. The International Date Line and the dawn of the new millennium. The Date Line no longer follows longitude 180°. The areas in daylight at 1543 GMT on 31st December are shown by the lighter grey. The Sun is rising on Millenium Island where the date is 1st January, identical with that in Tarawa. On Pitt Island sunrise is due 17 minutes later, but the Sun has already risen on the Wilkes Coast of Antarctica, where the date is 1st January, regardless of the modification of the date line. (If the black dots had been drawn to scale the tiny islands would be invisible on the figure.)

The line then changes direction towards the area of sea known as the Bering Strait. It reaches 169° west at 65.5 ° north, allowing the eastern extremity of Russia to be west of the International Date Line like the rest of the country.

An even more drastic diversion of the International Date Line was created as recently as 1st January 1995, after Kiribati had been extended eastwards by the inclusion of the Line Islands. Kiribati is a wide and watery political entity near the equator. It contains three groups of islands with a total population around 87 000 but it enjoys economic rights over around three million square kilometres of the Pacific Ocean. One of the inhabited Line Islands, now named Kiritimati, achieved fame reluctantly in the 1950s as Christmas Island, the base for tests of British nuclear weapons. (The old and new names of this island sound more alike than they look, because the letter combination *ti* is pronounced like an *s* in words like Kiribati and Kiritimati.) The capital of Kiribati is on the island of Tarawa, in the most westerly group, sometimes known as the Gilbert Islands. Tarawa and its neighbours lie north of New Zealand and are around 173° east, with clocks set twelve hours ahead of Greenwich Mean Time. The Phoenix Islands and the Line Islands are much farther to the east and have clocks set one or two hours ahead of those in the capital. In order to make the dates and days of the week the same throughout Kiribati, the International Date Line now makes a huge loop to the east. This means that clocks in the Line Islands are fourteen hours ahead of Greenwich Mean Time instead of ten hours behind.

A small uninhabited coral atoll previously known as Caroline Island is at a longitude just over 150° west but is now to the west of the International Date Line. This island was renamed Millenium Island because it was due to experience sunrise on 1st January 2000 at 0543 local time, when the time in Greenwich was only 1543 on 31st December 1999. The Kiribati claim to be winner of the race into the new millennium was not greeted with enthusiasm or acceptance by other countries wishing to claim that part of their territory would be the first to experience the dawn of the new millennium.

The first inhabited land to see the sunrise on 1st January 2000 was reckoned to have been Pitt Island in the group known as the Chatham Islands, east of New Zealand. Pitt Island is located near 176° west, but the 7.5° eastward shift of the International Date Line described earlier means that the island is west of the line. Although Pitt Island is about 26° farther west than Millenium Island, it is also about 35° farther south and so enjoys much longer periods of daylight during the summer of the southern hemisphere. The sunrise there was at 1600 on 31st December in Greenwich Mean Time, equivalent to 0545 on 1st January in local time. The discrepancy in the minutes shows that the local clocks align their minute hands with the clocks of Nepal.

Actually the piece of land with the first sunrise on 1st January 2000 was not in the Pacific Ocean. Much of Antarctica was enjoying continuous daylight, which means no sunrise, but the Wilkes Coast lies just outside the Antarctic Circle. At latitude 66° south and longitude 136° east, sunrise was at 0008 local time or 1508 Greenwich Mean Time, more than half an hour ahead of sunrise at either Millenium Island or Pitt Island. This Antarctic location is a long way west of 180°, so the win was achieved without recourse to any adjustment of the International Date Line. Nevertheless tourism and revelry on the Wilkes Coast were discouraged by the Antarctic Treaty, by the low temperature and often cloudy skies and by the distance from the nearest well stocked bar.

5

LIGHT AND THE ATMOSPHERE

5.1 Scattered light and twilight

The interactions between sunlight and the Earth's atmosphere are both useful and aesthetically pleasing. The extra daylight due to bending of the light in the atmosphere has already been mentioned in chapter 2. A more obvious effect of the atmosphere is the scattering of sunlight, which colours the sky and provides indirect illumination. For an astronaut on the surface of the Moon, the Sun shines out of a black sky. When it sets there is an abrupt transition from daylight to darkness and no twilight, because the Moon has virtually no atmosphere. Mars has some atmosphere, but it is very different from that on Earth. The Martian pressure is much lower and the major gas present is carbon dioxide. In addition, vigorous winds stir up a fine dust containing iron oxide. These particles absorb light at the short wavelength end of the visible spectrum, and reflect light of longer wavelength in all directions, giving the Martian sky a yellowish brown colour sometimes described as resembling butterscotch. (This was only established after some initial misinterpretations of images from the Viking Landers of 1976; the colour was later confirmed by the Mars Pathfinder.)

In the Earth's atmosphere changes in the direction of light occur even when clouds are absent. The effect known as *Rayleigh scattering* arises from variations in the scattering medium on a scale smaller than the wavelength of light. Lord Rayleigh was the first person to explain the scattering and calculate the way it

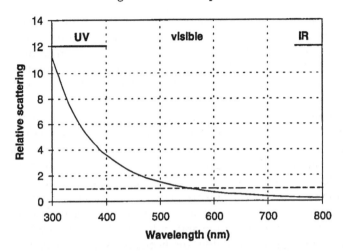

Figure 5.1. Wavelength dependence of Rayleigh scattering. The scattering rate for light of different colours is shown relative to that of green light with a wavelength of 550 nm, which is in the middle of the visible spectrum. Across the range of visible wavelengths the scattering rate changes by a factor of more than ten. The scattering rate is very high for ultraviolet and low for infrared.

depends on wavelength. In mathematical terms, the rate of scattering is inversely proportional to the fourth power of the wavelength. Figure 5.1 shows that this relationship implies a large change of the scattering rate over the wavelength range from 300 to 800 nm. Most of the ultraviolet light that avoids absorption by ozone in the upper atmosphere, and much of the violet and blue light, undergo scattering in a cloudless sky, which appears blue for most of the day. Most of the longer wavelength radiation reaches us directly, giving the Sun a yellowish colour. When the Sun is close to the horizon the path length through the atmosphere is extended so that the longer wavelengths also suffer appreciable scattering, shifting the colour of the Sun and of the sky around it towards red. Figure 5.2 (colour plate) is an example of the delightful colours that can be seen in skies around sunrise and sunset.

The colours of the sky at different times of day are obvious, but there is another manifestation of Rayleigh scattering that is easily overlooked. Nearly all the ultraviolet light that causes the damage to skin known as sunburn reaches the Earth's surface indirectly. A small umbrella provides shelter from direct sunlight and blocks most of the longer wavelength radiation including the infrared. The cool shade, however, can lead people to underestimate their exposure to ultraviolet, which arrives from all parts of the sky.

When the atmosphere contains particles larger than the wavelength of light, another process, known as *Mie scattering*, occurs. This is almost independent of wavelength, so that clouds usually appear white or grey. The same phenomenon is responsible for twilight. There are three formal definitions of twilight, all of which are based simply on the distance the centre of the Sun has reached below the horizon. *Civil twilight* is formally defined as beginning in the morning and ending in the evening when the centre of the Sun is 6° below the horizon. *Nautical twilight* lasts longer, as it begins or ends when the centre of the Sun is 12° below the horizon, which most people would regard as complete darkness. Astronomers are particularly fussy about scattered sunlight in the night sky, and try to avoid making observations during the period known as *astronomical twilight*, when the Sun is less than 18° below the horizon.

The duration of twilight depends on latitude and time of year. Figure 5.3 illustrates the way the twilight becomes longer as the latitude increases and the angle at which the path of the Sun crosses the horizon decreases. Twilight is briefest at the Equator, because the path of the Sun crosses the horizon at 90°, maximizing the speed of the Sun's descent or ascent through the region just below the horizon. Nautical twilight always lasts for less than an hour at the Equator, and is slightly shorter at the equinoxes than at the solstices.

Outside the tropics, the seasonal variations have a different pattern, and the duration of twilight is at its longest around the summer solstice. Whenever the latitude exceeds 54.5° (which is equal to 90° minus 23.5° minus 12°), the Sun at the summer solstice is never more than 12° below the horizon, implying that nautical twilight continues from sunset to sunrise. This is the case for the whole of Scotland and also the part of England north of Darlington. The variation in the duration of nautical twilight at 56°, roughly the latitude of Edinburgh, is illustrated and explained in figure 5.4. A noteworthy but rarely mentioned feature is that the

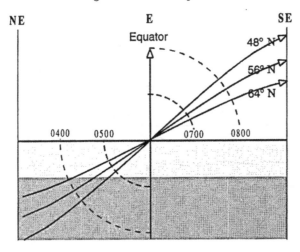

Figure 5.3. Nautical twilight duration at different latitudes. The course of the Sun, before and after sunrise at an equinox, is shown for the equator and for 48°, 56° and 64° N, which are near to the latitudes of Vienna, Edinburgh and Reykjavík. Time contours show local solar time. Nautical twilight exists when the Sun is in the pale grey zone, less than 12° below the horizon. In Reykjavík twilight lasts more than twice as long as on the Equator. The atmospheric refraction effect for the Sun near the horizon is about equal to the thickness of the black line representing the horizon.

shortest duration of twilight occurs just *before* the spring equinox and just *after* the autumn equinox.

Motorists in Britain are not required to know about the seasonal variations of the length of twilight. According to the Road Vehicles Lighting Regulations of 1989, front and rear lights are compulsory for moving vehicles between the times of local sunset and sunrise. In addition, headlights are compulsory during the hours of darkness, which are simply defined as starting half an hour after the local sunset and ending half an hour before the local sunrise. This thirty-minute period is independent of latitude and season and is always shorter than civil twilight within the United Kingdom.

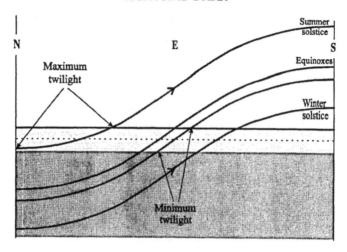

Figure 5.4. Nautical twilight near Edinburgh at different times of year. The black horizontal line represents the eastern half of the horizon. The course of the Sun for half a day, between local midnight and noon, is shown for latitude 56° N at various times of year. Nautical twilight exists when the Sun is in the pale grey zone, less than 12° below the horizon. Around the summer solstice nautical twilight extends throughout the night. Nautical twilight is shortest about two weeks before the spring equinox and two weeks after the autumn equinox, when the steepest part of the Sun's course is 6° below the horizon.

5.2 Polarization of light

As a gentle and unconventional introduction to the subject of light polarization, it may be helpful to consider grains of rice lying on a table travelling upwards in a lift. The direction of motion is vertical but the grains point in any horizontal direction. If the rice has been deposited casually, the grains would have random orientations. With a bit of effort (or a table with a cunningly designed surface) it is possible to align the grains partially or completely in any desired horizontal direction. The alignment of the grains of rice provides a crude but comprehensible analogy with the polarization of light.

Light can be described as an electromagnetic wave, which implies that it involves oscillating electric and magnetic fields.

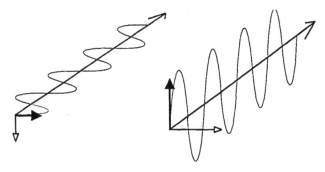

Figure 5.5. Light polarized in different directions. Two polarized rays are propagating in the same direction away from the reader. The oscillating electric fields are in different orientations, shown by the arrows with bold black heads. On the left the oscillating electric field is horizontal and the light is described as horizontally polarized. On the right the oscillating electric field is vertical and the light is described as vertically polarized. The oscillating magnetic fields are not drawn, but the arrows with white heads show their orientations.

These are normally at 90° to the direction in which the light is travelling and at 90° to each other, so the orientation of one defines the orientation of the other. Figure 5.5 shows two waves propagating in the same direction, but distinguished by different orientations of the electric field, analogous to arranging the ascending rice grains north-south and east-west.

Unpolarized light consists of a mixture of waves with electric fields pointing at random in all the possible directions. Such light is produced by many sources, including the surface of the Sun and tungsten-filament bulbs. Other sources, such as the semiconductor lasers that we shall meet in chapter 9, generate light with the electric field in a particular orientation. However, there are several ways in which polarized light can be produced by selecting one particular electric field orientation from an unpolarized source.

When light encounters a surface, the amount reflected depends on a number of factors, including the direction of polarization. As can be seen from figure 5.6, the reflectance is independent of polarization only when the light is incident at 90° or at 0°. At an angle known as the Brewster angle, the reflectance drops to zero

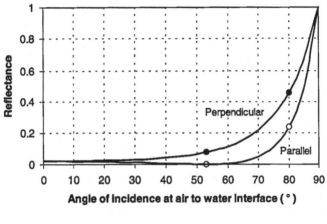

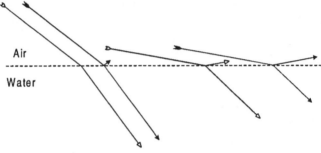

Figure 5.6. Refraction and reflection of light at a flat water surface. The fractions refracted or reflected when a ray meets any interface depend on the angle, on the polarization direction and on the refractive indices. In the lower diagram the length of the arrow indicates the intensity. At the Brewster angle (53° from vertical in air and 37° in water) light with polarization parallel to the page is not reflected at all (lower left). The circles in the upper diagram correspond to the incident rays in the lower one.

for one polarization direction, so the reflection consists entirely of light with the other polarization. In the next section we shall see that the selective reflectivity at the surface of a raindrop influences the polarization of rainbow light.

In some materials the arrangement of the atoms leads to bulk optical properties that can be exploited to select or adjust the po-

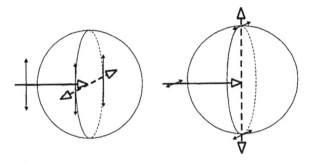

Figure 5.7. Effect of polarization on Rayleigh scattering at 90° to the incident light. The large unbroken arrows show polarized incident beams propagating from left to right. The small double-headed arrows show different orientations of the associated oscillating electric field. On the left this field is in the plane of the page, whereas on the right it is at 90° to the page. The large broken arrows show light that has altered direction by 90°. The polarization of the incident beam determines the new direction of propagation because the orientation of the electric field must be at 90° to both the original and the new direction. The spheres are drawn only to clarify the three-dimensional geometry.

larization direction. A classic example is the mineral known either as *calcite* or *Iceland spar*, a form of calcium carbonate. It has two refractive indices rather than one, so that an unpolarized ray entering the crystal is split into two rays polarized at right angles to one another and following different paths. This naturally occurring material is transparent to both polarizations and produces a directional separation of the light, whereas other substances achieve separation by absorbing light of one polarization while transmitting the other.

In 1928 Edwin Land, then a student at Harvard, invented the first synthetic material of this type. Polaroid J Sheet was a plastics material containing tiny parallel crystals of quinine sulphate periodate. In 1938 Land produced a significantly better material, known as Polaroid H Sheet. The molecular alignment caused light of one polarization to be absorbed, while light polarized in the other direction was transmitted with a very low scattering loss. A wide range of materials has subsequently been developed,

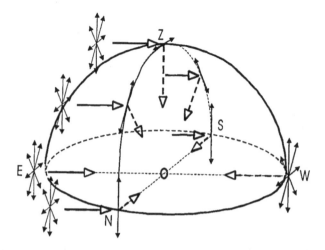

Figure 5.8. Direction-dependent polarization of scattered sunlight. For an observer at O the sky resembles a hemisphere. When the Sun is just above the eastern horizon, direct unpolarized sunlight propagates westward as shown by the large unbroken arrows. The orientation of oscillating electric field, shown by the small double-headed arrows, is not restricted to a unique direction. Scattered sunlight reaching the observer from various parts of the sky is shown by large broken arrows. Back-scattered light from the western sky has undergone a 180° change of direction and remains unpolarized. Light scattered towards the observer from the semicircular arc from north to south via the zenith Z has undergone a 90° change of direction and so is polarized in a direction that is a tangent to this arc.

allowing optical engineers to choose a Polaroid material that is optimized for a particular wavelength range.

Light from the Sun is unpolarized while it travels through space. However, Rayleigh scattering in the atmosphere not only changes the direction in which light travels, but also acts as a rather effective polarization selector when the change in direction is 90°. Light scattering occurs only if the scattered light retains the original orientation of the oscillating electric field. This is illustrated in figure 5.7, in which two incident beams with different polarization directions undergo scattering and change direction by 90°. Light polarized in the vertical direction before scattering

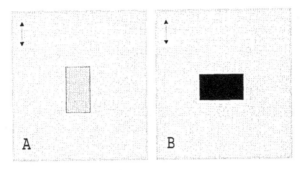

Figure 5.9. Polarized sky light seen through filters with different orientations. Scattered light from the northern or southern sky shown in figure 5.7 has a vertical polarization, indicated here by the black arrows. The small rectangles represent polarization-selective filters in front of the observer's eyes. In A the orientation of the filter allows the vertically polarized light through. In B the orientation of the filter ensures that light with this polarization is blocked, thereby making the sky appear dark.

can retain this polarization only if it propagates horizontally after scattering, whereas light polarized in the horizontal direction is restricted to propagation upwards or downwards. No polarization selectivity arises when the change of direction is 180° and the scattered light follows its original path in the reverse direction.

The polarization of scattered light from a clear sky is illustrated in figure 5.8. It shows that light from the sky is unpolarized when it comes from the direction of the Sun (in this diagram from the east or the left) or from the opposite direction (from the west or the right), but is polarized when it comes from an arc across the sky at 90° to the direction of the Sun. The polarized nature of light reaching an observer from any part of this arc is very easy to detect by looking at the sky through a sheet of Polaroid or some other polarization-selective filter. When the filter is rotated, the transmission of the light from the sky changes and the sky appears alternately darker and lighter as the rotation continues. This is illustrated in figure 5.9.

Many living creatures can identify the direction of polarization of scattered sunlight and exploit it for navigational purposes. It is less widely known that the human eye has a vestigial ability to distinguish the polarization direction. This topic is discussed in chapter 6.

5.3 Rainbows

Almost two hundred years ago William Wordsworth wrote: 'My heart leaps up when I behold a rainbow in the sky'. These simple words neatly express the delight that this most colourful of atmospheric light patterns evokes. Rainbows can be enjoyed without any understanding of the underlying optical effects, but some knowledge of their characteristics is not hard to acquire.

Rainbows are visible when drops of water in front of you are illuminated by light directly from the Sun in a cloudless part of the sky behind you. The drops are usually rain, but the same effects arise in spray from waterfalls or hoses. The rays from the Sun undergo a sequence of interactions with different parts of the curved surface of each drop. Each time the surface is encountered, the propagation direction changes, some light being reflected while the majority is transmitted at an angle dependent on the ratio of the refractive indices of water and air. The refractive index of the water is 1.332 for red light, increasing to 1.343 for violet light. Consequently, the different colours follow different paths after entering the drop, red light being deviated least.

The behaviour of light encountering a spherical water drop is illustrated in figure 5.10. The lengths of the arrows provide a rough indication of the intensity of the light leaving in various directions. Most of the light that enters the drop emerges at the next encounter with the surface, but a small fraction is reflected back inside the raindrop. The upper part of figure 5.10 shows how a primary rainbow is produced by light that has undergone one internal reflection before making its exit. Looking at a rainbow, you see light of different wavelengths from a multitude of drops simultaneously, with the red light mainly from drops in an arc in a direction about 42.5° from the antisolar point, i.e., the direction exactly opposite the Sun. The violet light comes from drops 2° closer to the antisolar direction. If the Sun is more than 42.5° above the horizon, the antisolar point is so far below the opposite horizon that no primary rainbow can be seen. Expressed in everyday terms, you cannot see rainbows when the Sun is high –

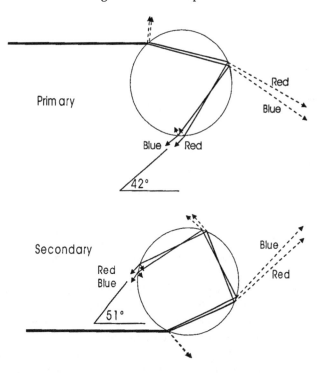

Figure 5.10. Refraction and reflection of sunlight in a raindrop. At each interaction with the surface of the drop, most of the light is refracted and the rest is reflected. The differences between red and blue light have been slightly exaggerated for greater clarity. The production of the primary rainbow involves one internal reflection, the red (long wavelength) light emerging at a steeper angle than the blue (short wavelength) light. The secondary rainbow results from two internal reflections, so it is weaker and has the colours reversed.

unless you contrive to be above the raindrops. From an aircraft you can sometimes see the entire circle of a rainbow below, with the shadow of the aircraft at its centre.

If the sunlight is bright and the background dark, a weak secondary rainbow is visible outside the primary. The lower part of figure 5.10 shows how this rainbow is produced by light that has undergone two internal reflections. The extra surface encounter

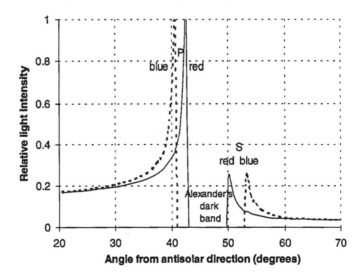

Figure 5.11. Angular dependence of the intensity of light redirected by raindrops. With the Sun directly behind and the rain in front of the observer, the pure red outside edge of the bright primary rainbow P is seen as an arc roughly 42.5° from the point directly opposite the Sun. The secondary rainbow S is less intense and has a pure red inside edge as an arc at about 50°. Between the two red arcs is Alexander's dark band, from which no light redirected by rain reaches the observer. Inside the primary rainbow there is so much redirected light of all colours that the sky is conspicuously lighter.

reverses the colour sequence and makes the secondary rainbow fainter and broader. Most red light in the secondary rainbow comes from drops about 50° from the antisolar point.

Figure 5.11 shows the angular distribution of red and blue light in both rainbows. At one side of each peak the light intensity falls abruptly to zero. Between the two abrupt red edges there is a band about 7° wide that contains no light redirected by the raindrops toward the observer. This is known as *Alexander's dark band*, because around 200 AD the Greek philosopher Alexander of Aphrodisias wrote the earliest known document describing it. On the other side of each peak the intensity of the redirected light falls only slowly, so the non-red parts of both rainbows include some light with wavelengths longer than the predominant one. At low or high angles there is a mixture of all visible wavelengths at almost the same low intensity. This lightens the colour of the sky inside the primary rainbow and (to a lesser degree) the colour of the sky outside the secondary rainbow. The resulting appearance of the sky in the vicinity of a rainbow is illustrated in figure 5.12 (colour plate).

A noteworthy but far from obvious feature of rainbows is that they consist of polarized light. When a ray of light encounters a surface between two materials the proportions of light reflected and refracted depend on the direction of the oscillating electric field and on the angle of incidence. When the ray meets the surface at angles close to zero or to 90°, the reflectivity is hardly affected by the direction of polarization. As shown in figure 5.6, at intermediate angles the reflectivity depends on the polarization direction, being least when the oscillating electric field lies parallel to the plane containing both the incoming and the outgoing rays. At the Brewster angle, the reflectivity drops to zero for the parallel polarization only. The Brewster angle can be calculated from the refractive indices of the two materials on opposite sides of the surface. It is about 53° for the air-to-water interface and 37° for the water-to-air interface. Although the angles involved in the raindrop do not match the Brewster angle exactly, most of the light with parallel polarization passes through the raindrops without being reflected. The creation of the rainbows is therefore almost entirely due to reflection of light polarized at 90° to the page containing figure 5.10.

This means that any rainbow consists almost entirely of light with the polarization aligned in the direction tangential to the arc, although this is not readily apparent to the unaided eye. If Polaroid sunglasses are worn, however, only vertically polarized light can reach the eyes. Consequently the horizontal top of the rainbow cannot be seen, while the vertically polarized light from the sides of the rainbow is not only visible but more impressive against the darker background. If the head and the Polaroid sunglasses are tilted by 90°, only the top of the rainbow is visible. The viewing of rainbows through a polarization-selective material also enhances the contrast between the different areas of grey sky adjacent to the rainbows and makes Alexander's dark band more conspicuous.

5.4 Cloudy skies

Clouds contain water droplets so small that they fall very slowly due to air resistance, and can be kept aloft by rising air. Droplets with a diameter between 0.02 and 0.2 mm are considerably larger than the wavelength of light and scatter light by Mie scattering. This process is virtually independent of wavelength, so clouds normally appear white or grey, with the grey becoming darker as the cloud becomes thicker or denser.

The intensity of light reaching the ground from a totally cloudy sky depends on a number of variables, especially the height of the sun and the thickness of the cloud cover. Architects need a standard that is typical of a cloudy sky when they design buildings and calculate the window areas required to achieve a specific amount of internal illumination by daylight. In Britain and other countries with a temperate climate the usual standard for such calculations is the CIE Overcast Sky. CIE is the abbreviation of *Commission Internationale de l'Éclairage*, an organization concerned with illumination standards, which covers diverse topics including colour photography, traffic lights and the protection of valuable paintings from fading. In spite of the French name, the central office is in Vienna.

The CIE Overcast Sky is equivalent to a hemispherical shell emitting diffuse light non-uniformly and producing a total illuminance of 5000 luxes (lx) at the centre of the circular base. Illuminance is formally defined as the luminous flux density at a surface, but in everyday terms it is the total amount of light reaching a surface from every direction. The assessment of illuminance involves physiology as well as physics, because it takes into account the unequal response of the average human eye to different colours or wavelengths, a topic that is discussed in chapter 6. Though IR and UV radiation may be present, they are invisible and so contribute nothing to illuminance. An illuminance of 5000 lx can be appreciated by observing or imagining a horizontal surface illuminated only by a 150 W tungsten-filament bulb 200 mm directly above it.

The brightness of the CIE Overcast Sky changes from horizon to zenith in the manner illustrated in figure 5.13. The varying brightness is a fairly good match to the experimental data for real cloudy skies and is proportional to the value of a simple mathe-

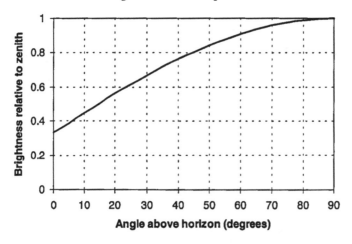

Figure 5.13. Brightness of overcast sky. The CIE Overcast Sky is a standard sky matching the typical characteristics of a sky entirely filled with cloud. The sky directly above the observer is three times as bright as the sky near the horizon.

matical function with the form $(1 + 2\sin\theta)/3$ where θ is the angle above the horizon. The important point is simply that this cloudy sky is three times as bright at the zenith ($\sin\theta = 1$) as it is at the horizon ($\sin\theta = 0$).

5.5 Halos

When the atmosphere is at a temperature below $0\,°C$, condensation of water vapour forms ice directly without passing through a liquid form. At low altitudes the crystals of ice usually acquire large and complex forms and fall as snowflakes. At higher altitudes the crystals often remain so tiny that they fall very slowly and may even maintain their altitude for a considerable time in uprising air currents. Such ice crystals occur in the cirrostratus clouds, which typically form about 6000 metres above sea level at a temperature around $-20\,°C$. They produce optical effects known as halos when they interact with light arriving directly from the Sun or sometimes from the Moon. Only the simplest and most frequently observed phenomena will be described here.

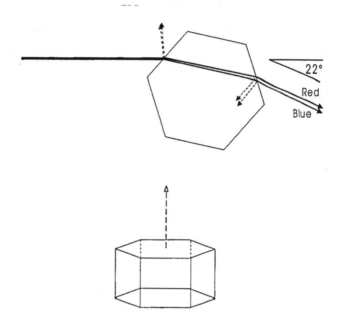

Figure 5.14. Refraction of sunlight in a hexagonal ice crystal. Light passing through two non-parallel and non-adjacent faces of the upper ice crystal undergoes a change in direction by an angle around 22°, slightly less at the red end of the spectrum and slightly more at the blue end. Randomly oriented ice crystals create a bright circular halo with an angular radius of 22° around the Sun. If most ice crystals in the cloud adopt the preferred orientation with the axis of the hexagon vertical, as shown in the lower part, they create bright patches on the left and right of the Sun.

Whereas snowflakes have intricate shapes based on hexagonal patterns, the tiny ice crystals normally exist as simple hexagonal prisms. When light passes through such a crystal, it encounters two surfaces with a precise angle between them. When the two surfaces are parallel the emerging ray is parallel to the incoming ray, but when the angle between the two surfaces is 60° the light changes direction, as illustrated in figure 5.14. The interaction between light and ice crystals has some features in common with the interaction between light and raindrops, but there are also some noteworthy differences.

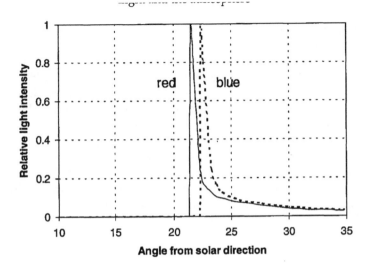

Figure 5.15. Angular dependence of the intensity of light redirected by ice crystals. Small hexagonal ice crystals in the upper atmosphere produce a halo around the Sun. The halo has an abrupt and pure red inside edge at 21.5° from the direction of the Sun. The angular separation of red and blue is much less than that occurring in rainbows.

The light that produces the circular halo has been deviated by passing through two non-parallel surfaces, but has not undergone any internal reflection. The resultant change in direction is close to 22°, so that the halo appears as a ring around the Sun. (This angle is roughly the same as the maximum separation of the tips of the thumb and the little finger with an outstretched arm.) Figure 5.15 illustrates the characteristics of the circular halo for light at opposite ends of the visible spectrum. It shows that the angle separating the colours is only about 0.8°, which is much less than the 2° and 3° in the primary and secondary rainbows shown in figure 5.11, which has different scales. The rate of change of refractive index with wavelength is virtually the same in ice as in liquid water, but in a raindrop the effect of the change is amplified by the internal reflection at different parts of the curved surface. It is even more difficult to see the colour separation in a lunar halo, because vision at night relies on rods, which are not colour-sensitive (but the colour separation can be confirmed by photography).

For a halo to be visible as a complete and uniform circle around the Sun, ice crystals in many orientations are needed, but the aerodynamics of falling ice crystals may sometimes lead to certain orientations being preferred, so that some parts of the circular halo are more conspicuous than others. When most of the ice crystals have their hexagonal axis vertical and the Sun is low in the sky, as it inevitably is in Arctic and Antarctic regions, bright patches can be seen in the sky 22° to the left and right of the Sun. Such a patch is known as a *parhelion* or, more often, as a 'sundog'. A halo and its associated sundogs are shown in figure 5.16 (colour plate).

6

SEEING THE LIGHT

6.1 The human eye

The detection of light patterns involves a sequence of processes not only in the eye but also in the brain. The interpretation of the data transmitted from the retina to the brain is the last but by no means the least part of the sequence. However, it is best to begin by considering only the structure of the human eye, which is shown in simplified form in figure 6.1.

The adult eyeball is almost spherical and has a diameter of about 24 mm. The eyeball of a newborn baby has already grown to almost three-quarters of the adult size, a feature contributing to the baby's attractiveness as well as its ability to see. The transparent, slightly bulging front surface of the eye, the *cornea*, is actually a five-layer structure. The curvature of the surface, and the refractive index change from 1.0003 in air to about 1.34 inside the eye, bend the incoming rays of light so that they become convergent. Additional convergence is produced by the flexible lens, which has an even higher refractive index, cunningly varying from about 1.41 at the centre to about 1.39 at the edge. The cornea contributes about 70 per cent of the total convergence and the lens about 30 per cent. The combination is able to focus an image of the scene on the light-sensitive retina at the rear of the eyeball. To adjust the focus for viewing objects at different distances, the shape of the flexible lens is modified by the action of the *ciliary muscle*. This adjustment is known as *accommodation*.

The intensity of light reaching the back of the eye is controlled by the *iris*, which contains opaque pigments. The diameter of the aperture or *pupil* can be changed in a fraction of a second from

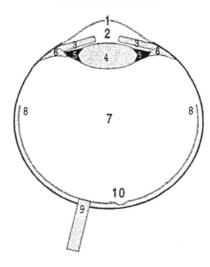

Figure 6.1. Horizontal cross section of the right eye. Light enters through the cornea (1) and passes through the aqueous humour (2), the lens (4) and the vitreous humour (7) before being absorbed by photoreceptors in the retina (8). There is a blind spot (9) on the retina where the optic nerve leaves the eye. The sharpest vision and best colour discrimination occur at the fovea (10). The size of the pupil and hence the amount of light reaching the retina is controlled by expansion or contraction of the iris (3). The lens shape is altered by the action of the ciliary muscle (6) via the ligaments (5) in order to optimize the focus for objects at different distances.

8 mm to 2 mm by the actions of two layers of muscle fibres in the iris. In the forward layer the fibres are arranged around a circle so that their contraction reduces the diameter of the pupil, whereas the fibres in the rear layer are arranged radially so that their contraction increases the diameter. The ratio (16:1) between the maximum and minimum pupil areas is much less than the range of illumination levels that are encountered. Consequently, an impression of the ambient brightness is maintained in spite of the adjustment of the pupil diameter. In primates and other large animals with binocular vision, the sizes of the two pupils remain more or less equal, even when one eye is illuminated more than

the other. Curiously, the pupils are always small during sleep, although most sleep occurs during darkness.

At the back of the eye lies the delicate light-sensitive membrane called the *retina*. It contains several layers, the details of which need not concern us here. Although you might expect the light-sensitive cells called *photoreceptors* to be on the front (or inner) surface of the retina, they are actually near the back. To reach them, the light has to find its way past the complex network of nerves. All types of photoreceptor convert the energy provided by the light into electrical signals that eventually reach the visual cortex at the rear of the brain. The retina covers such a large area that the human eye has a remarkably wide field of view. On one side of the eye it is the projecting nose and not the absence of photoreceptors that limits the field. On the other side the field extends more than 100° from straight ahead. Although the periphery of the retina is unable to distinguish detail, it is valuable for survival. This is just as true in the Motorway Age as it was in the Stone Age because a driver can look at the road and the vehicles ahead and nevertheless be aware of a vehicle alongside.

Where the optic nerve leaves the eye, there are no photoreceptors. The resulting blind spot on the retina can be seen just left of the centre in figure 6.1. It always lies between the nose and the centre of the retina. Because the image on the retina is inverted, there is a small invisible region within the visual field, located about 14° to the right of centre for the right eye. For the left eye it is 14° to the left, which means that the images of an object cannot fall on the blind spots of both eyes simultaneously. Even with one eye closed the blind spot generally goes unnoticed, because the brain assumes that the invisible region resembles the neighbouring areas. You can check this with the help of figure 6.2.

One effect of increasing age is that the lens becomes less flexible, making it gradually more difficult for a normal sighted person to focus on close objects. Figure 6.3 shows that the flexibility and the associated accommodation decline steadily throughout adult life. The deterioration only becomes really conspicuous when objects must be more than 300 mm away to be seen clearly, which implies that the flexible lens cannot increase its power by more than 3 dioptres. This situation is typically reached in the late forties, when people with normal vision begin to need glasses with convex lenses for close work.

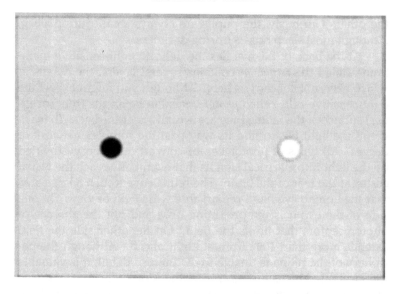

Figure 6.2. Demonstration of the blind spot on the retina. Close your left eye and stare at the black dot with your right eye. As you reduce the distance between your head and the page, you can find a position in which the white dot appears to be masked by the grey area. Alternatively close your right eye, stare at the white dot and alter the distance to make the black dot disappear.

In addition the lens becomes less transparent with increasing age. Coagulation of protein from old cells produces a general loss of transmission through the lens, which may eventually become almost opaque, a condition described as a cataract. Even in the absence of a drastic transmission loss across the entire visible spectrum, there is less transmission at short wavelengths so that the amount of violet light reaching the retina is reduced and the image may be perceived as yellowish and dingy. The shortest wavelength detectable by the eye at any age is primarily determined by the optical transmission characteristics of the flexible lens. Consequently, people who have undergone surgery to remove the defective natural lens acquire an extended visual range and can detect light in the near UV that was invisible previously.

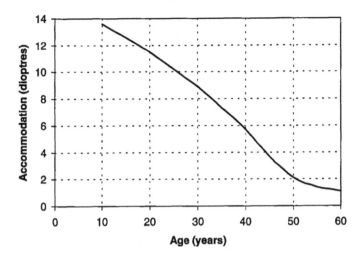

Figure 6.3. Effect of age on the flexibility of the lens in the human eye. The accommodation describes the range of adjustment of the flexible lens, allowing an object to be kept in focus as it becomes closer. The dioptre is a unit describing the strength of a lens and is the reciprocal of the focal length in metres.

The longest detectable wavelength is determined by the characteristics of the photoreceptors. Four types, with different light-sensitive pigments, are present in most human eyes. Figure 6.4 shows their distribution within the retina.

At low levels of illumination, vision depends on pencil-shaped receptors called rods, containing a monomolecular layer of the pigment *rhodopsin*. This pigment is bleached by exposure to light and has its maximum sensitivity at a wavelength just over 500 nm. Because bluish green light is strongly absorbed but red and violet are not, rhodopsin has a purple appearance and hence is also known as *visual purple*.

There are over 120 million rods, distributed over a wide, almost hemispherical region at the rear of the eyeball. Each rod is extremely sensitive, being able to respond almost to a single photon. Particularly near the edges of the retina, many rods may share a common nerve to the brain, so ability to work at low light intensity is achieved at the expense of spatial resolution. This

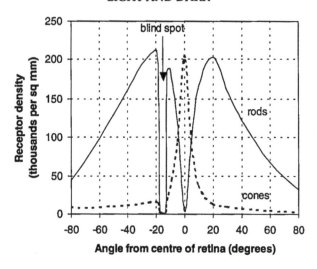

Figure 6.4. Distribution of receptors across human retina. The density of receptors is shown for a horizontal cross section through the right eye. The cones are concentrated near the centre of the retina, whereas the rods are widely distributed and rare at the centre. Both types of receptor are absent from an area about 2 mm in diameter where the optic nerve exits from the eye, creating a blind spot on the nasal side of the retina.

form of vision, usually called scotopic vision, provides a wide field of view and high sensitivity, but cannot distinguish colour or resolve fine detail. It is virtually impossible to read a book by moonlight.

In daylight, saturation of scotopic vision occurs. The rods become ineffective as the replacement rate for rhodopsin fails to match the rate of consumption. It takes about thirty minutes in darkness to restore the full sensitivity of rod vision. The synthesis of rhodopsin within the body requires vitamin A or its precursors, so a diet deficient in these materials leads to 'night blindness'. It is said that when airborne radar systems were introduced during the Second World War, producing a conspicuous increase in the effectiveness of allied night fighter aircraft, the existence of the new technology was kept secret by ascribing the improvement to carrots and cod-liver oil consumed by the pilots.

Under bright conditions the dominant process is *photopic* vision, based on cone-shaped receptors that have less than a hundredth of the sensitivity of the rods. There are about 6.5 million cones, mostly concentrated within a small area, known as the *fovea*, located near the centre of the retina, although some are spread at low density over an area wide enough to provide some photopic vision over a wide angle.

The centre of the fovea is rich in cones and free of rods, thereby producing the highest visual acuity. Here the cones are tightly packed at intervals of only a few micrometres (μm), thereby achieving a maximum density of around 200 000 per square millimetre. Using this part of the retina, a normal eye can resolve the detail of an object to around a sixtieth of a degree (one minute of arc). This is defined as a visual acuity of 1 (sometimes expressed as a fraction such as 6/6 or 20/20). In less technical terms, this means that an accidental nudge of a golf ball by one millimetre when lining up a putt could be noticed by another player less than 3.5 metres away. Diffraction of the light as it passes through the front of the eye is probably as important as the spacing of the cones in determining the ultimate resolution of the eye, at least when bright light produces small pupil apertures. Resolution is even better for extended objects such as wires. Under the most favourable conditions, the human eye can detect a black wire only 0.1 mm thick against a white background from a distance of 20 m, corresponding to an angle no greater than one second of arc (0.0028°).

6.2 Colour vision and colour blindness

6.2.1 Colour vision

In the normal human eye there are three types of cone, distinguished by the presence of one of three iodopsin pigments. The three pigments differ in their sensitivity to various wavelengths of light within the visible spectrum. Sensitivity implies that light of the corresponding wavelength is absorbed and has its energy converted to an electrical signal in a nerve.

Figure 6.5 shows that the peak sensitivities of the three types of cone are at wavelengths around 565, 530 and 425 nm. The cones associated with the peak at the longest wavelength are usually

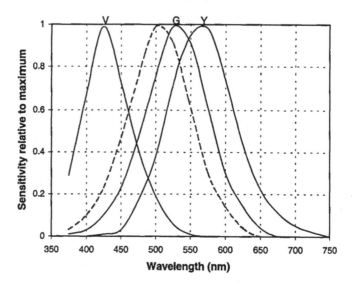

Figure 6.5. Wavelength dependence of response of four types of photoreceptor. The response of rods, shown by the broken black line, has a maximum around 505 nm. The responses of the three types of cone are unequally spaced with maxima around 425 nm (violet), 530 nm (green) and 565 nm (yellow). Each curve is normalized and does not take into account the unequal numbers of photoreceptors or their relative sensitivities.

described as 'red-sensitive', even though the peak at 565 nm actually lies in a part of the spectrum generally accepted as yellow. (The familiar orange-yellow light from sodium-vapour streetlamps has a wavelength around 589 nm, which is 24 nm farther towards the red.) The peak at 420 nm is in the violet part of the spectrum near the short wavelength limit of vision, but the associated cones are usually called 'blue-sensitive'. Only the green-sensitive cones are accurately named. (In this section the traditional nomenclature is, however, retained for all three types.) The red-sensitive cones are the most numerous. The blue-sensitive cones comprise less than ten per cent of the total cone population in an average human eye and are even scarcer at the centre of the fovea, but compensate for their small numbers by having a higher sensitivity.

Each pigment has an appreciable response to wavelengths on either side of its peak sensitivity. For the cones with peak sensitivities at 565 or 530 nm, some response occurs across more than 70 per cent of the visible spectrum. The blue-sensitive cones have a response range that extends from the near UV to green. However, the shortest wavelength response is not exploited because the lens is opaque to UV radiation. The cumulative effect of UV absorption by the lens hastens the development of cataracts.

One function of the visual cortex is to interpret the cone responses in terms of hue. For light of wavelength around 650 nm, which is generally perceived as red, the stimulation of the green-sensitive cones is considerably smaller than the stimulation of the red-sensitive cones, but it is by no means negligible. The unequal spacing of the three peaks and the unequal numbers of the three cone types may benefit the overall performance of the human eye because clarity is probably more important than colour assessment. It may be advantageous to have few blue-sensitive cones in the fovea and to rely mainly on two types of cone having maximum responses at wavelengths only 35 nm apart, less than ten per cent of the total visible range. Because the refractive indices of the materials in the eyeball vary with wavelength, it is impossible to achieve perfect focus of the image on the retina for all wavelengths simultaneously. Relying mainly on a narrow band of wavelengths at the centre of the total visible range allows a sharper image to be conveyed from the retina to the brain. The wavelength-dependence of refractive index can be experienced during routine sight tests. It is a standard procedure for optometrists to present black letters on green and red backgrounds alternately in order to ascertain precisely which corrective lens produces the best compromise for the eye under test.

Another outcome of the unequal wavelength spacing of the three cone types is that the human eye can detect differences in the colours of two monochromatic sources much better in the middle of the visible spectrum. Yellowish greens stimulate the red-sensitive and the green-sensitive cones almost equally. In this part of the spectrum two monochromatic wavelengths only 1 nm apart can be distinguished. At the extremities of the spectrum only one type of cone is being stimulated significantly and so a wavelength difference less than 10 nm may be hard to detect. As either end of the visible range is approached, the colours appear to get darker but the hue hardly changes.

At 390 and 750 nm the total sensitivity of the retina is only around 0.01 per cent of the maximum value. These wavelengths may be regarded as the normal limits for human optical response, though the cut-off point depends on the light intensity and is not abrupt. The ratio between the longest and shortest wavelengths detectable by the eye is almost 2:1. In the realm of music or acoustics, such a wavelength ratio would be described as a range of one octave, which is tiny compared with the range of acoustic wavelengths detectable by human ears, which may extend over ten octaves – equivalent to a wavelength ratio of more than 1000:1. In passing, it is worth noting that the emission peaks of the three phosphors used to generate coloured pictures on TV and computer screens do not match the sensitivity peaks of the cones. These phosphors emit red, green and blue light, with wavelength peaks that are roughly equally spaced, for example at 610, 550 and 470 nm. Experiments have shown that such a combination of wavelengths produces a wide colour range.

6.2.2 Colour blindness

The ability to create a wide variety of colours by mixing three primary colours, and the existence of colour blindness, have long been recognized, but it was only in 1802 that a coherent explanation of colour blindness emerged. The British doctor Thomas Young was interested in the colour blindness of John Dalton, the chemist who pioneered the atomic theory. (Dalton's eyes are still preserved in Manchester.) Young correctly identified the cause as the absence of one of three colour-selective sensors normally present. ('Daltonism' is a term still sometimes used to describe colour blindness, especially the most common forms in which colours in the red-to-green part of the spectrum are confused.) Difficulty in distinguishing reds and greens occurs in about 7.7 per cent of Caucasian men (1 in 13), while the incidence in Caucasian women is about 0.6 per cent (1 in 13^2). This type of colour blindness is less common in other ethnic groups.

Although it was known at the time that red–green colour blindness was concentrated in certain families and was more common in men than women, it was not until the early years of the 20th century that the underlying genetic mechanisms became clear. The genes responsible for conveying the ability to create the red-sensitive and the green-sensitive cones are both located on the same chromosome, the X-chromosome. Furthermore, their DNA sequences have a large amount in common. The differentiation between these two types of gene appears to be recent in evolutionary terms, having developed less than 40 million years ago. The closeness of the locations of these two genes and the similarity in their structures increase the probability of errors during the formation of cells for sexual reproduction. As a result of such process errors, some X-chromosomes have acquired extra copies of the gene that produces green-sensitive cones at the expense of other X-chromosomes that possess none. Extra copies of this gene have little effect on human colour perception, but total absence of the gene for green-sensitive cones produces the most common form of colour blindness, known as *deuteranopia*. Figure 6.5 shows that in the absence of green-sensitive cones, light with wavelengths in the range 550 to 750 nm stimulates only one type of cone. This makes it virtually impossible to differentiate between colours in the green-to-red part of the spectrum. Monochromatic light with a wavelength around 485 nm stimulates both blue-sensitive and red-sensitive cones to a similar extent, making it difficult to distinguish from grey, though it would appear bluish green to people blessed with all three types of cone.

Within a Caucasian population the gene for green-sensitive cones is lacking in about 1 in 13 of the X-chromosomes. Males inherit an X-chromosome from their mother and a Y-chromosome from their father. With only one X-chromosome, the incidence of the deficient X-chromosome directly determines the incidence of deuteranopia in males. Females inherit two X chromosomes (one from each parent) and it is only when both X chromosomes lack the gene for green-sensitive cones that this form of colour blindness occurs. Consequently deuteranopia is much less common in females.

The lack of the gene for the red-sensitive cones is known as *protanopia*. It is less common than deuteranopia, but it is also associated with the X-chromosome and therefore more common in males than females. Colour blindness associated with the absence of blue-sensitive receptors is known as *tritanopia*, a very rare defect due to a missing gene on chromosome 7. Because this chromosome has no connection with sexual characteristics, the incidence of tritanopia is similar in males and females. The DNA sequence that codes for blue-sensitive receptors has developed separately for about 500 million years and has little in common with the sequences for red-sensitive or green-sensitive receptors.

A few people have no cone function at all and see only shades of grey. This condition is known as *achromatopsia*. The condition it is extremely rare worldwide, but it occurs in around one person in twelve on Pingelap, one of the Caroline Islands in the western Pacific. The gene responsible is present in almost 30 per cent of the population of this island, but because the gene is recessive, achromatopsia is manifest only in people inheriting the gene from both parents.

Where it may be important, colour schemes are sometimes devised with the commoner forms of colour blindness in mind. Many computer software packages, such as Word or Excel, use blue and yellow more liberally than red and green. Website designers are encouraged to use labels and variations in texture instead of relying on variations in colour. Traffic lights have red at the top and green at the bottom, so that their message can be deduced from position as well as from colour. The colour coding in mains cables has been chosen so that colour-blind people can identify the three leads correctly.

6.3 Polarization sensitivity

We saw in chapter 5 that skylight becomes polarized if it has undergone a 90° change in direction. Some birds and insects have eyes that can identify the direction of polarization. They make use of this impressive ability for the purpose of navigation. Human beings normally employ some artificial aid, such as a sheet of Polaroid. Nevertheless in some circumstances the unaided human eye does have the ability to identify the polarization characteristics of light.

Although its existence has been known since 1846, this visual capability is not widely known and the underlying mechanism remains poorly understood. The effect has the strange name of *Haidinger's brush*, because it was described first by the Austrian scientist Wilhelm Haidinger. (The surname should be pronounced like 'hide singer' without the first letter of the second word.) It is generally ascribed to a small patch with polarization-dependent properties near the centre of the retina.

You can observe the effect with only rudimentary equipment and a little practice. The first requirement is a uniform, well-lit white surface. A blank white screen of an active VDU is particularly effective. The second requirement is a method of selecting linearly polarized light. Polaroid sunglasses are fine, but they cannot be worn in the usual manner because a rapid change of their orientation is needed. With one eye closed or covered, view the white area through the polarization-selective filter. At intervals of a few seconds, rotate the filter rapidly through 90° to change the direction of the polarization of the light reaching the eye.

Immediately after a change of the polarization direction, a faint mustard yellow pattern may be perceived on the white background near the centre of your field of vision. The form, illustrated in figure 6.6 (colour plate), has been described as a double-ended brush, an hourglass or a bow tie. The long axis of the yellowish shape lies perpendicular to the electric vector of the polarized light. For some observers, there may also be a bluish purple region aligned in the orthogonal direction, producing an effect like two bow ties of different hues forming a Maltese cross. The effect lasts only a few seconds, but it can be regenerated in the opposite orientation by switching the polarization direction through 90° again or (less effectively) by removing the filter. The transient nature of this phenomenon may be because the human brain treats the colour pattern as anomalous, in much the same way as the brain compensates for the blind spot on the retina.

6.4 Speed of response

Projected movies consist of sequences of still pictures with brief dark intervals between them. Because the retina and its associated nerves do not respond immediately to changes and require time to recover after stimulation by light, the brain interprets this sequence as steady motion and fails to perceive the dark intervals. The recovery period is dependent on the light intensity, because cones and rods have different recovery characteristics. For rods, recovery is slow. For cones, when there is sufficient light for them to be activated, recovery takes around 50 ms ($\frac{1}{20}$ s). Cine cameras usually operate at 24 fps (frames per second). During projection, the shutter closes twice for every frame, once to conceal the film movement to the next frame and once in the middle of each stationary period. This means that the screen is illuminated 48 times per second, with each frame projected twice. The result is a flicker-free impression of steady movement.

In the early years of cinematography (and still in some 8 mm home movies), frame rates were slower, namely 16 fps. Today it is not uncommon for early film to be presented inappropriately at 24 fps, producing a mildly comic effect either deliberately or through incompetence. Projection at 16 fps with three shutter closures per frame is not hard to arrange and produces flicker-free movement at the correct speed.

Films need not be projected at the same speed as they were recorded. When the action is too slow to retain attention, for example the growth of a plant or the motion and development of a cloud, it is easy for the frames to be recorded less frequently. This is known as time-lapse cinematography. Conversely, when motion is very rapid, for example a hummingbird's wings, frames can be recorded at a high rate. Projection at the standard rate enables the action to be seen in slow motion. In German there is a neat term, *eine Zeitlupe*, which could be translated literally as a time lens.

The time delays involved in the photochemical and electrochemical processes that contribute to vision are not fixed, but depend on the intensity of the illumination. This can be illustrated by observing an oscillating object with one eye behind clear glass and the other behind dark glass so that the intensity of the light reaching the eyes is far from equal. Sunglasses with dark glass for one eye only are a rarity, so the visual cortex has to interpret inputs of an unfamiliar sort. Figure 6.7 shows a plan view of an easily constructed arrangement sometimes known as *Pulfrich's pendulum*.

A convenient oscillation frequency of about one cycle per second can be obtained with a small heavy object suspended on a 250-millimetre cotton thread. A suitable amplitude for the oscillation is around 10° either side of the vertical or just under 50 millimetres either side of the central position. Neither the time nor the amplitude need to be set precisely, but the oscillation must be

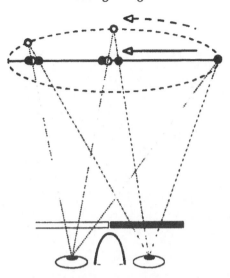

Figure 6.7. Effect of different optical response on perceived position. The black circles represent an object oscillating from right to left and back again along the straight line. If the light entering one eye is made less intense than the light entering the other, the resulting signals from photoreceptors in the two eyes reach the brain after different delays. At a particular instant, the brain receives information from the right eye about a grey position and from the left eye about a white position. Consequently the distance is judged incorrectly and the object appears to follow an elliptical track.

confined within a left–right vertical plane without any movement towards or away from the observer. The movement should be observed with both eyes near the level of the moving object. No illusion occurs if light from the object reaches the left and right eyes with equal intensities. However, when the light intensities are unequal due to dark glass in front of one eye, the oscillating object appears to move towards and away from the observer as well as moving left and right, producing an illusion of an elliptical rather than a straight path. The direction around the illusory ellipse changes from anticlockwise to clockwise if the dark glass is moved from the right eye to the left one. The effect is more

Figure 1.6. The picture 'Christ before the High Priest' by Gerrit van Honthorst (1590–1656) provides a fine example of the chiaroscuro style. (© National Gallery, London.)

Figure 5.2. Midwinter dawn sky over Essex.

Figure 5.12. Rainbows over Alberta with contrasting shades of sky.

Figure 5.16. Winter sky over Iowa with a halo around the sun and a sundog on each side. (© Kip Ladage/Ladage Photography.)

Figure 6.6. The key to recognizing the optical effect called Haidinger's Brush.

Figure 7.1. Grant's zebra, a common subspecies of the plains zebra.

Figure 7.2. Chapman's zebra, another subspecies of the plains zebra.

Figure 7.3. Hartmann's zebra, with a white belly and a dewlap.

Figure 7.4. Grévy's zebra, with a white belly and many narrow stripes.

Figure 7.5. Zebra danio (horizontal stripes) and black-tailed humbug (vertical stripes).

Figure 7.10. Light emission from a female firefly or glow-worm (classified as a beetle). (© Raymond Blythe/Oxford Scientific Films.)

Figure 8.1. The Roman lighthouse at Dover was constructed around 43 AD, shortly after the invasion of Britain by the legions of Emperor Claudius.

Figure 8.6. The semaphore tower at Chatley Heath, 33 km SW of London, was a relay station for messages between the Admiralty and the naval dockyard at Portsmouth.

striking if the movement takes place among stationary objects that provide reference points for comparing distances.

The illusion arises because the time for the optical input to be converted to a nerve impulse and conveyed to the visual cortex of the brain becomes longer as the light becomes less bright. The information reaching the brain at any instant has come from two different positions of the moving object because of the unequal delays. Instead of perceiving two objects at different positions on the actual path, the brain reconciles the two directions by assigning the object a position farther or nearer than the true location.

6.5 Optical illusions

Earlier in this chapter we noted how the brain contrives to compensate for the presence of a blind spot on each retina, while the illusion created by the arrangement in figure 6.7 demonstrates how information from the retinas can be misinterpreted. These phenomena indicate that visual perception involves the brain as well as the eyes. For a deeper understanding of vision and the tricks it can play, you need to be aware of the paths along which information travels to the visual cortex at the back of the brain. The nerve fibres from the two eyes meet and divide again at a junction known as the *optic chiasma* before continuing to structures in the thalamus or midbrain. Some of these structures control reflexes, but the left and right *lateral geniculate nuclei*, usually abbreviated to LGNs, process information relating to colour, brightness and movement and send signals onward to the visual cortex. (The word 'geniculate' is absent from many dictionaries and is just an obscure description of the shape, which resembles a bent knee.) The surprising feature is that the left halves of both retinas are connected to the left LGN, while the right halves are connected to the right LGN. Consequently, the left LGN handles all the information from the right half of the visual field, while the scene on the left is entirely processed in the right LGN. Damage to one LGN or an adjacent nerve leads to loss of half the visual field, which can be a greater handicap than the loss of one eye. When the signals from the LGNs reach the visual cortex they are compared with those previously experienced in order to make sense of the scene.

118

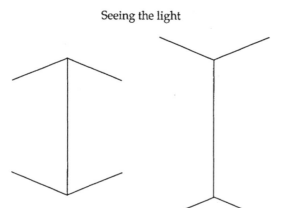

Figure 6.8. The Müller-Lyer illusion. In this well known illusion the two vertical lines have equal length. The apparent difference probably arises because the visual cortex cannot avoid interpreting the five-line 2D drawings as external and internal perspective views of a rectangular 3D object.

With some optical illusions it is possible to identify the part of the visual pathway where the misinterpretation occurs. The well-known *Müller-Lyer illusion,* shown in figure 6.8, is almost certainly produced in the visual cortex. People unfamiliar with perspective views of rectangular objects do not experience this illusion. For other illusions the cause is not so easily identified or explained. Figure 6.9 shows two patterns where a set of straight lines appear bent as the result of crossing another set of lines at varying angles. The illusion may originate in the LGNs overestimating small angles and sending incorrect information to the visual cortex.

Part C of figure 6.10 shows an illusion that is difficult to explain. It is often known as 'the café wall illusion' because it was noticed some years ago in the brickwork of a small café in Bristol. However, this illusion had already been discovered in the late 19th century. If certain conditions are fulfilled, as in part C of this figure, the lines separating the horizontal layers of bricks in a wall or squares in the figure appear tilted in alternating directions. This illusion depends on the position and on the darkness of the three components in the pattern. It requires the squares in successive layers to be offset and not aligned with squares either of the same type (producing straight vertical columns) or of the

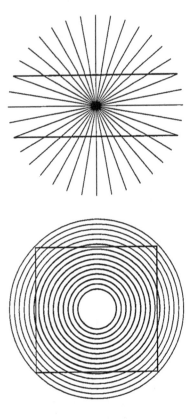

Figure 6.9. Bent line illusions. A straight line can appear curved if it crosses other lines at varying angles. The parallel lines in the upper diagram seem further apart at their centres than at their ends. The lines on opposite sides of the square in the lower diagram seem further apart at the corners than at the middle of the sides.

opposite type (producing a chess board). In addition the boundaries separating the squares in a manner akin to the mortar in a brick wall must have an appropriate thickness and be lighter than one type of square and darker than the other type. The latter criterion is not met in parts A, B and D of figure 6.10 and so the illusion is absent

An illusion of distorted straight lines can occur in a purely black-and-white pattern. Figure 6.11 shows how a simple

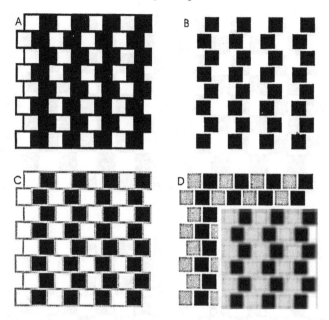

Figure 6.10. The café wall illusion. Each of the four patterns has the same geometrical design, based on squares in horizontal layers. The only difference between the patterns A, B and C is that the boundaries are black, white or grey. Pattern D is derived from pattern B by changing the white squares to grey. The layers appear undistorted in A, B and D, but the grey boundaries in C create the illusion that the layers are a pile of alternating wedges.

checkerboard pattern can be affected by superimposing numerous small white squares on the black ones. On a casual glance, the plain black squares appear larger than the black squares with four small white squares, so that rows and columns seem to have a non-uniform width. This may be related to the orientations of the two small white squares in the diagonally adjacent black squares.

Some illusions affect perception of colour or darkness instead of geometric form. If the visual cortex were simply recording the input from the retina, plain white paper would seem to be coloured differently in daylight and in candlelight, because the sun is much hotter and brighter and produces light with a far

Figure 6.11. Checker board distortion. The addition of certain patterns of small white squares on the black squares creates an illusory distortion of the horizontal and vertical lines. The plain black squares tend to appear larger than the decorated black squares.

greater blue-to-red ratio. With film designed for daylight, a photograph of a candlelit scene can look horribly orange. However the visual cortex compensates in an inconspicuous manner for the difference in brightness and colour balance, so that you perceive white paper as 'white' in spite of differences in illumination. The adjustment that creates such constancy of colour is normally useful but sometimes it occurs inappropriately. The next figures present illusions that are believed to arise in this way.

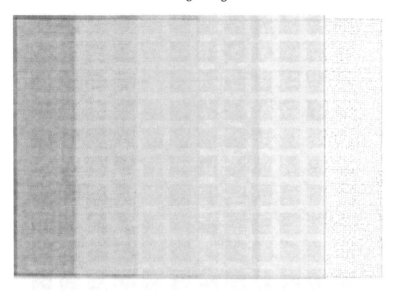

Figure 6.12. Colour uniformity distortion. Each of the six rectangular patches has a uniform grey shade, but the grey appears graded because one adjacent patch is paler and the other adjacent patch darker.

Figure 6.12 shows six patches, each of a different uniform shade of grey. Because each patch has two adjacent areas that are not equally dark, an impression of graded greyness is created where none exists. Figure 6.13 presents a rather more complex illusion in a black and white pattern known as *Hermann's grid*. The intersections of the white grid lines between the black squares appear pale grey. If the diagram is rotated through 45°, pale grey diagonal lines appear to be present.

The creation of a sensation of colour from a black-and-white pattern is particularly intriguing. The effect can be experienced by observing the rotation of *Benham's discs*, which are named after a supplier of toys exploiting this illusion in 1894. Less impressive versions of this class of illusion had already been around for more than fifty years. Figure 6.14 shows a simple but effective example. The disc is divided into black and white semicircular areas. On part of the white area there is a set of concentric black arcs. In this example all the arcs extend for the same angle of 60°, but

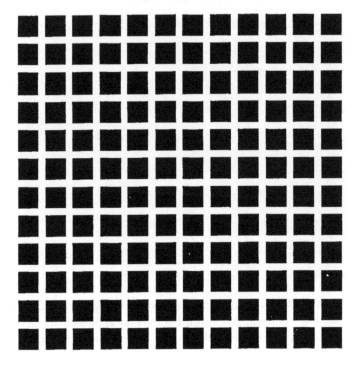

Figure 6.13. Hermann's grid. This figure is simply constructed from black squares on a white background. Nevertheless grey spots are usually perceived where the horizontal and vertical white lines cross. The effect is stronger at the edges of the visual field than at the centre. If the pattern is rotated by 45°, a grey grid at 45° to the white grid may be perceived.

they have different orientations relative to the boundary between the two semicircles. When the disc rotates at 50 or more revolutions per second, the alternation of black and white is so rapid that the semicircles cannot be distinguished and most of the disc appears pale grey. The black arcs increase the black-to-white ratio at certain distances from the centre and so are perceived as uniformly dark grey concentric circles. The grey pattern perceived with rapid rotation is what would be expected from an understanding of the speed at which the eye responds, the subject of the previous section.

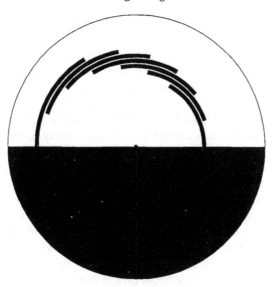

Figure 6.14. Benham's disc. When a disc of this type rotates rapidly about its centre, the human eye cannot distinguish the black and white parts. Most of the disc appears pale grey, while the black arcs are perceived as darker grey concentric circles. As the rotation rate is reduced, the circles seem to acquire a range of colours and resemble a grubby rainbow. The red appears at the inside with clockwise rotation and at the outside for anticlockwise rotation. The pale areas inside and outside the series of dark circles also acquire slightly different colours.

The surprise occurs when the rotation rate is reduced. Within the range four to ten revolutions per second, the dark circles generated by the black arcs appear coloured. The perceived colours depend on the angular position of the arc relative to the straight boundary between the semicircles. The regular arrangement of arcs shown in figure 6.14 produces a rainbow effect. The variation in perceived colour for the set of dark circles created by the arcs is conspicuous, but the two pale areas inside and outside also acquire slightly different hues. Reversal of the direction of rotation reverses the entire colour sequence. The colours become more striking as the light intensity increases, and they are more vivid in light from a tungsten filament lamp than in daylight. The latter effect is presumably related to the higher red-to-blue ratio in the light from the lamp.

Although this class of illusion has been recognized since the 19th century, no completely satisfactory explanation has yet emerged. The majority of hypotheses involve the different speeds at which the three types of cone react to and recover from stimulation by white light, the blue-sensitive cones being slowest. Nevertheless, the illusion cannot be ascribed solely to the different rates of the photochemical processes occurring within the cones, because contrasting faint false colours are also perceived in the inner and outer pale areas where the arcs have only an indirect effect. It seems that the visual cortex is being misled to compensate for something that does not actually require compensation. You might enjoy predicting and then observing the behaviour when the arcs are white on black instead of black on white.

It is not only scientists and toy makers who are interested in patterns that play strange tricks on the human eye and brain. Optical art, usually abbreviated to 'op art', is an abstract art form developed in the 1950s and 1960s. The works contain sharp-edged patterns of lines that are usually black and white, though sometimes coloured. They are designed to exploit optical phenomena in the eyes and brains of the observer that make the patterns appear to flicker, vibrate or wriggle.

Even a nominally stationary eye staring at a small object makes tiny and repetitive shifts of direction. These eye movements extend for only a fraction of a degree and include flicks, drifts and high frequency tremors, which evolved to produce a more effective stimulation of the photoreceptors in the retina. When the distance from the observer is such that the spacing of a pattern corresponds to an angle nearly matching that of the eye movements, there can be an interaction producing a kind of moiré effect. This creates a somewhat disturbing illusion that the pattern is in motion. The shape of the lines is less important than regularity and strong contrast, which means that analogous effects can be achieved with a range of patterns, of which there are two examples in figure 6.15. Sometimes the effect occurs after a prolonged stare, but on other occasions it arises after the pattern has been moved to alter the angle or the distance relative to the observer.

The pioneer of op art after the Second World War was Victor Vasarély who was born in Hungary but moved to Paris in 1930.

Figure 6.15. Illusion of movement created by patterns of black and white lines. When such patterns are viewed from an appropriate distance, interaction between the spacing and tiny eye movements creates an uncomfortable impression of motion.

The leading British exponent was Bridget Riley, who became a celebrity in the 1960s while still in her early thirties. When she arrived in New York in 1965 to make an appearance at an exhibition of her work at the Museum of Modern Art, she noticed with dismay that some of her designs seemed to have already been incorporated into clothes on sale in the shops. Although the public fascination with op art declined after a few years, she continues to be a productive and respected artist in a variety of abstract styles. Some of her paintings, such as *To a Summer's Day* (1980), use a range of colours but earlier ones, including *Movement in Squares* (1961) and *Fall* (1963), show what can be achieved in black and white. The Tate Modern in London has several examples of her work, but unfortunately they are sometimes in store rather than on display.

7

ZOOLOGICAL DIVERSIONS

7.1 Colour vision in animals

The ability of animals to perceive colour varies greatly from species to species. Nocturnally active mammals tend to have poor colour discrimination, because their visual systems rely mainly on rods for light detection. The majority of mammals active during daytime are dichromatic, meaning that they have only two cone pigments, one with a response peak in the yellow-to-green part of the spectrum and the other in the blue-to-violet range. Consequently their colour discrimination is probably similar to that of people with the commonest form of colour blindness. Waving yellow rags at bulls might be more provocative and dangerous than waving red ones, because yellow light is closer to the peak sensitivity of the commonest cones. In a china shop, though, such animals should be able to discriminate between cheap terracotta flower vases and expensive Wedgwood Blue Jasper ware.

The only mammals known to possess trichromatic vision are among the primates. Monkey species native to Asia and Africa are generally trichromatic as humans are, whereas monkeys native to South America normally have dichromatic vision. One species of monkey, *Callithrix jacchus jacchus*, from South America, does have retinal characteristics that may be regarded as a stage in the development of full trichromatic vision. Each male has only two types of cone pigment, one for the blue and violet end of the spectrum and the other with a peak response at one of three longer wavelengths (543, 556 or 563 nm). This means that the males are all dichromatic, but they fall into three groups, each group having a slightly different colour perception. Some females are dichromatic and can be classified in the same three groups as the males. However other females are trichromatic, with three types of cone pigment comprising any two of the three longer wavelength forms in addition to the short wavelength form. Such a distribution can be explained by the presence on every X-chromosome of a gene for one of three pigments with peak sensitivity at 543, 556 or 563 nm. Males have one Y-chromosome and one X-chromosome and so possess only a single gene for a long-wavelength pigment. Females have two X-chromosomes, which may contain two identical or two dissimilar genes for long-wavelength pigment. Because the majority of individuals of this species have dichromatic vision, trichromatic vision could be regarded as a deviation from the norm. Although the trichromatic females may be better at finding fruit amongst foliage and therefore be able to produce and feed more offspring, they have no way of passing the trichromatic vision on to male descendants, unless an improbable reshuffle of DNA occurs so that the two dissimilar pigment genes become located on a single X-chromosome instead of two. Such a change may have been the mechanism by which Old World primates acquired trichromatic vision.

Although we tend to regard primates as advanced forms of life, in terms of colour vision many birds, fish and other animals surpass them. Most birds that are active during daytime are endowed with four or five cone pigments, which provide a wider and richer experience of colour than that enjoyed by any mammal. One of these pigments responds to light at UV wavelengths that are invisible to us. The plumage of a raven appears black to humans, but it reflects UV radiation, so birds would see a raven's plumage as bright, not black at all.

The male and female forms of the blue tit *Parus caeruleus* appear very similar to humans but to each other they do not look the same at all. The blue crown on the head makes it easy for us to recognize the species but not the sex. For the blue tits themselves there is no such problem because in the UV the male and female crowns are quite distinct. The females show a preference for mates with crowns possessing the brightest UV colouration. Swedish zoologists have shown that the sex life of popular males can be ruined by the application to females of UV-blocking materials of the type used for preventing sunburn in human skins.

Bees are another example of creatures able to see UV, but at the expense of response to red. Their eyes respond to light with wavelengths within the range 300 to 650 nm. Flowers photographed using filters that absorb red and transmit UV can appear remarkably different, and give us some idea of what a bee might see. For really advanced colour perception, it is hard to beat the mantis shrimp, which has no fewer than ten types of pigment in its photoreceptors. It would be interesting to know whether its brain has the capacity to exploit fully such richness of colour sensation. Perhaps those mathematicians who are accustomed to thinking in ten-dimensional hyperspace are the only people with a chance of comprehending the profusion of colour enjoyed by the mantis shrimp.

7.2 Zebras

7.2.1 Species and subspecies

Zebras are wild animals belonging to the same genus, *Equus*, as horses and donkeys. Hundreds of years ago they could be found in many parts of Africa, but hunting reduced their populations and agriculture has taken some of the land in which zebras used to roam. The influx of European settlers with firearms caused a rapid decline in numbers during the period between 1850 and 1950, and some subspecies were annihilated. In the last few decades, however, conservation measures in Africa and breeding in zoos have improved the chances of survival for the remaining endangered types.

For many people a zebra is just a zebra, but zoologists distinguish three species, two of which are divided into subspecies. Although all three species can interbreed and even breed with donkeys, the species remain distinct because the crossbred offspring are infertile. The black and white stripes have conspicuously different characteristics in the three species, so that it is not difficult to learn to identify them. Within any one species the pattern varies slightly from one animal to the next, but recognition of individual animals is best left to the specialist (or to the zebras themselves). The most numerous and widespread species is the plains zebra *Equus burchelli*. These animals thrive over a wide range of habitats in Africa, even at 4000 metres above sea level.

Several subspecies are recognized, the commonest being Grant's zebra *Equus burchelli boehmi* with a simple alternating pattern of black and white, illustrated in figure 7.1 (colour plate). Another subspecies of the plains zebra is shown in figure 7.2 (colour plate). Chapman's zebra *Equus burchelli antiquorum* resembles Grant's zebra in many respects, but its broad white stripes have darker midlines, known as 'shadow stripes'. Two other subspecies of the plains zebra, *Equus burchelli burchelli* and *Equus burchelli quagga*, were hunted to extinction by European settlers. The last sighting of a quagga in the wild was in 1878, and the last zoo specimen died about five years later. Both these unfortunate subspecies had stripes at the front end only.

Another species, the mountain zebra *Equus zebra*, was once widespread in southern Africa. Having narrowly escaped extinction through hunting, these animals now enjoy protection in a national park in the Republic of South Africa. The mountain zebra is slightly smaller than the plains zebra, but the most conspicuous differences from the plains zebra are the pronounced dewlap under the throat, the plain white belly, the very broad stripes on the side of the rump and the fine herring-bone pattern of narrow transverse black stripes on top of the rump where the tail emerges. The species contains two subspecies known as Hartmann's zebra *Equus zebra hartmanni*, shown in figure 7.3 (colour plate), and the cape zebra *Equus zebra zebra*, which is the smallest of all the zebras.

Figure 7.4 (colour plate) shows the third species *Equus grevyi*, known as Grévy's zebra or the imperial zebra. It is larger than the other zebras and is often considered to be the most handsome. A few thousand years ago its range extended northwards as far as the mouth of the Nile, but now it can be found only in Ethiopia, Somalia and Kenya. The stripes are narrower and more numerous than those on the other species; this is particularly conspicuous on the rump. As with the mountain zebra, the stripes do not extend underneath the body. There is a broad black stripe lengthwise above the spine. Its call is different from that of other species and resembles the braying of a donkey.

The variation of the number, width and distribution of stripes on different species has been ascribed to the triggering of the stripe development mechanism at different stages of pregnancy. According to this theory, each stripe is about 0.4 millimetres wide when it first appears on the foetus, but it subsequently grows at the same rate as the developing part beneath it.

It is less easy to explain how the black and white stripes on zebras increase the chance of survival. Camouflage is an unconvincing explanation, as the form and the high contrast of the pattern do not match the background of the open plains where the animals graze. The simple tawny colour of lions would appear to be a much more effective camouflage. Another hypothesis is that the small variations of the stripe pattern serve to identify individuals within a herd of these gregarious animals; but wild horses identify other members of their herd satisfactorily without the aid of stripes. A plausible suggestion is that the patterns make the zebra more difficult to catch because predators suffer some kind of visual disturbance akin to that produced by figure 6.15. Such an effect is hard to confirm, because later natural selection would probably have favoured the survival of predators unaffected by such a disturbance. At present the benefit zebras might gain from their striped patterns remains a mystery.

7.2.2 Other zebra-striped animals

The black and white stripes of zebras are so striking to a human observer that the name has often been applied to other species and even to inanimate objects. Zebras have provided the descriptive name for British pedestrian crossings and French speakers use the term *zébrure* to describe striped patterns. The zebra finch *Taeniopygia guttata* has only a tenuous link to the zebra. It is only the tail feathers of this Australian bird that have black and white stripes.

There are more than forty species of fish with English names that include the word zebra. Such nomenclature is easily justified when the fish has vertical stripes with almost equal widths for black and for white, such as the zebra seabream *Diplodus cervinus cervinus*, which lives in the warmer parts of the eastern Atlantic Ocean. However, naming of species after the zebra is rather haphazard, as figure 7.5 (colour plate) shows. The fish *Brachydanio rerio* in the upper half of this figure is known either as the zebra fish or as the zebra danio, but the colour pattern is more like that of a chipmunk. Parts of the body are brown and the lengthwise stripes have only a moderate contrast. Conversely, the black-tailed humbug *Dascyllus melanurus*, shown in the lower half of figure 7.5, has a striking pattern of vertical stripes, but gets its English name from the black-and-white sweet.

7.3   Piebald coats and unusual goats

Many mammals have coats that are black and white, but the evolutionary advantage of a particular pattern often remains a mystery. In many cases patches of black and white seem to be there to make the animal more visible and are the reverse of camouflage. The badger *Meles meles* can be found in many parts of Europe and is well known for the conspicuous black and white stripes on its face. Less well known features of its colouring include white tips on the ears and the greyish colour of the body due to hairs that change colour along their length. We shall return to such patterns on individual hairs later.

An animal that exploits the visibility of a black and white pattern is the ring-tailed lemur *Lemur catta*, found in southern Madagascar and illustrated in figure 7.6. The tail contributes more than half of the total length of the animal and has about thirteen zones of each colour. When the ring-tailed lemur moves around the forest floor during the day, its tail is held vertical, making it as conspicuous as possible. This makes it easier for family groups to stay together.

In some species the black and white regions of the body are few in number and large in area. The giant panda *Ailuropoda melanoleuca*, shown figure 7.7, is a familiar example. The second half of the giant panda's Latin name is derived from the Greek words for black and white. The panda has so much visual appeal that it was chosen as the symbol of the World Wildlife Fund (WWF). But the cuddly image is inconsistent with its antisocial lifestyle. A giant panda tries to keep other pandas away from its patch of bamboo forest, a valuable resource for an animal that spends most of its waking life supplying a rather inefficient digestive system. Although giant pandas have rather poor eyesight, the stark black and white contrast ensures that they become aware of each other's presence. Most people remember the white face with a black nose, round black ears and areas of black around each eye, but the body hair also has particular features, to which toymakers pay less attention. The black hair on the front legs and feet

Figure 7.6. Ring-tailed lemur. The tail of this inhabitant of Madagascan forests is longer than its body and has probably evolved to make the animal visible to other members of its group.

Figure 7.7. Giant panda. The dark ears and dark patches around the eyes make this animal appealing to human beings; but the contrasting areas of black and white probably evolved to make it easier for these solitary animals to keep apart from each other.

continues across the shoulders and in a narrow strip with distinct edges over the back. The boundary between the white body and the black back legs is sometimes less abrupt. Another unusual anatomical feature is that the front feet have five almost parallel claws supplemented by a modified wrist bone functioning like a human thumb. Zoologists have disputed for many years whether giant pandas should be classified in the bear family, in the raccoon family or in a family containing only the one species. At present classification as bears seems to be preferred.

Even simpler arrangements of black and white can be found on other wild animals. A simple three-zone black-white-black sequence is characteristic of another endangered species, the Malayan tapir *Tapinus indicus*. The forward black area includes the head, neck, shoulders and front legs. Most of the long body is white but the rear, including the back legs, reverts to black.

In some animals the black-white-black sequence has not arisen naturally but has been selected by human intervention in the breeding process. An example is the breed of cattle known as the Belted Galloway or Beltie, shown in figure 7.8. This breed is thought to have originated in southwest Scotland in the 16th century. These handsome animals are noted for their lean meat and for the absence of horns. They are also very hardy, a characteristic that may be ascribed to natural selection in the Scottish climate. The most obvious external feature is the white belt around the middle of the body, occupying from 20 to 40 per cent of the body length. The precisely located colouring of the Belted Galloway may be contrasted with the random black and white markings of the Friesian, known as the Holstein in the USA.

Switzerland is often considered as the centre of European goat breeding, with seven domestic breeds being recognized. In the northwest of the country, where French is the main language, the most numerous goats are white and belong to the Saanen breed. In Ticino, the canton in the south where Italian is the main language, most goats belong to the Nera Verzasca breed. (*Nera* is the Italian for black and Verzasca is the name of a valley to the north of Lake Maggiore.)

The most intriguing breed of Swiss goat gives the impression that the animal has been assembled from two half-goats of contrasting colours. Appropriately, this breed originated in a Swiss canton that contains a distinct language boundary. In the east-

Figure 7.8. Belted Galloway cow. The belted Galloway is a handsome, hardy, hornless breed, primarily reared for beef. It originated in Scotland, but during the last sixty years it has become increasingly popular in Australia, New Zealand, Canada and the USA. The black and white version is most common but there is also a brown and white version of the breed.

ern half German predominates, the name of the canton is *Wallis* and the river flowing westward is *der Rotten*. In the western half French predominates, the name of the canton is *Valais* and the river is *le Rhône*. There are at least two thousand goats in this canton that display their own distinct transition, the entire front half being black and the back half being white. Figure 7.9 shows the abrupt boundary between the two zones. The names *Walliser Schwarzhalsziege* or *valaisan à col noir* become 'Valais blackneck goat' in English, but the black area in fact includes the head, shoulders and front legs.

The mountain village of Zermatt is famous for its proximity to the Matterhorn, but it also offers a close view of these hairy goats every morning and afternoon during the summer when they walk nonchalantly along the car-free main street on their way to and from their mountain pastures. The goat breeders hold shows and markets in the spring for males and in the autumn for females. At these times the goats have their long hair washed and combed. In the selection of prize-winners and breeding stock, much attention is given to the position, straightness and abruptness of the boundary between black and white around the entire

Figure 7.9. Valais blackneck goats. In the Swiss canton of Valais many goats have a black front and a white rear. Adults of both sexes have shaggy coats and beards. Only goats with an abrupt colour change half way along the body are selected for producing the next generation.

circumference of the body. For females, the external appearance is considered more important than the milk yield. Males not chosen for fathering the next generation have a short life expectancy. The eye-catching appearance of the breed can thus be ascribed to *un*natural selection.

7.4 Jellicle cats are black and white

These words of T S Eliot describe only a fraction of the population of domestic cats. The coats of *Felis catus* can have a vast range of colours and patterns. Although the genetic mutations responsible have arisen naturally, many types survive only because cat breeders have actively favoured animals with pretty or unusual appearances. In the wild, natural selection would have eliminated new types of cat that were unable to obtain sufficient food and mates. Skin and fur colours are due to melanins, which are pigments produced in skin cells known as melanocytes. The dark colours are produced by eumelanin, whereas phaenomelanin is responsible for the orange colour described by breeders as red. White fur is free of melanins, owing to the action of genes that hinder the migration of melanocytes through the body of the developing embryo.

In the cells of a cat there are nineteen pairs of chromosomes, four pairs fewer than in a human cell. The cells for reproduction contain nineteen unpaired chromosomes, so that the fusion of spermatozoon with ovum produces a new cell in which one chromosome of each pair is derived from each parent. As in human beings, one pair determines the sex, females possessing two X-chromosomes and males one X and one Y. There are several genes determining the characteristics of the fur, and they are not all located on the same chromosome. The gene responsible for orange fur is located on the X chromosome, which is why this colour of fur does not occur with the same frequency in male and female cats.

A convenient starting point is a consideration of the genes that must have contributed to the fur characteristics of the first cats to become domesticated. It is generally accepted that all such cats were originally tabby with a pattern of black and brown stripes, probably similar to the wild cat *Felis silvestris grampia* in Scotland today. The stripes formed a fairly regular sequence of bands around the body, the legs and the tail. Such a pattern may have provided camouflage useful for stalking prey. The common domestic shorthair cats described as mackerel tabby still display this pattern, which arises from the presence of several independent and dominant genes.

One gene determines the presence of 'agouti' characteristics. This term, derived from the name of a small South American rodent, describes the colour pattern of each hair in the lighter parts of tabby fur. Viewed from afar such fur simply looks brown, but close examination shows that each agouti hair changes colour along its length. If you blow gently at the lighter fur of a cooperative tabby cat to reveal the full extent of the long hairs as well as their tips, it becomes apparent that each hair changes from black to beige and back again. Breeders call such hairs 'ticked'. Without the presence of the agouti gene, these hairs would be uniformly dark and indistinguishable from hairs in the darker areas of the tabby pattern. Two other genes are responsible for the hue of the dark pigment and its concentration in the dark regions of the tabby pattern. Yet another gene determines the form of the tabby pattern. There is also a non-orange gene, an apparently perverse concept that I shall return to later.

The occurrence of cats with black fur all over is thought to have been one of the first conspicuous changes. It was caused by a mutation of the agouti gene to a non-agouti gene. The new type of gene is recessive, meaning that a cat with an agouti gene from one parent and a non-agouti gene from the other has the same external tabby characteristics as a cat with two agouti genes. A black cat has no agouti fur because it has two non-agouti genes, one inherited from each parent. Nevertheless a black cat does not necessarily have black parents. The mating of two tabby cats, each having one agouti and one non-agouti gene, leads to kittens with varied genetic inheritances that may produce different fur characteristics. Each kitten inherits half the genes of each parent, but it is chance that selects which genes are passed on and which are discarded. Any kitten of these parents has a one in four chance of being black due to two non-agouti genes, a two in four chance of being tabby with one gene of each type and a one in four chance of being tabby with two agouti genes.

The precise form of the tabby pattern is determined by a patterning gene that has undergone mutations. In one extreme version of the pattern, the growth of black hair is suppressed and all the fur is ticked. This mutation is co-dominant with the basic stripe pattern, whereas most other pattern modifications are due to recessive genes, and only manifest themselves in cats that have inherited genes for the modified pattern from both parents. What breeders call 'the classic tabby pattern' has irregular marbled stripes.

The presence of orange fur arose from a mutation of the non-orange gene on the X-chromosome into the orange gene. The effect of the orange gene is to replace dark colours with orange and to suppress the action of any non-agouti gene. Consequently the orange colour is never uniformly dark but is always associated with some kind of pattern, although the contrast between the two shades of orange may be slight. Because male cats have only one X-chromosome, the occurrence of the orange gene in a large population equals the fraction of male cats that have orange fur. This is commonly around 30 per cent in Britain. On female cats the equivalent orange coat occurs only when the non-orange gene is absent, which requires both X-chromosomes to carry the orange gene. The fraction of such females is 30 per cent of 30 per cent, which equals 9 per cent. The incidence of orange fur on male cats is analogous to that of red–green colour blindness in men, discussed in chapter 6, as the genes responsible are located on the X-chromosome in both cases. Female cats with no orange gene on either X-chromosome form 70 per cent of 70 per cent, i.e. 49 per cent of the female population. The remaining 42 per cent of the female population have the orange gene on one X-chromosome and the non-orange gene on the other. Instead of one gene being dominant and the other recessive, both genes contribute to the colour. Patches of orange fur and of dark fur occur side by side, and the cat is described as a tortoiseshell or 'tortie'. The combination of these two colours, with or without additional white patches, is found on female cats only. The very rare male tortie has two X chromosomes and one Y, and is invariably sterile.

White fur is free from pigment, and is the result of any of three different genetic mechanisms. The albino white and dominant white genes both produce totally white coats, but the gene responsible can be identified from eye colour, pink or blue. The commonest cause of white fur is the white spotting gene, which produces white patches that can comprise any proportion of the coat. The areas of white fur have well defined boundaries and can co-exist with any of the common fur types: tabby, black, orange and tortie. The gene that promotes white areas does so by opposing the migration of melanocytes through the body of the developing embryo. The effectiveness of this opposition is variable, and the areas of white and dark fur often appear to have been selected arbitrarily. In former President Clinton's cat Socks only the white paws remained free of melanocytes. At the other end of the scale, the fur is mostly white with black, orange or tabby patches that are usually on the head, back and tail. As a general rule, white fur becomes more common as the distance from the spine increases. The tail is the part that is least likely to be covered in white fur, so a white tail usually belongs to a totally white cat. Siamese cats have a gene that allows the formation of melanocytes only on the coolest parts of the body, namely the muzzle, ears, tail and legs.

7.5 Cephalopods

The class of marine molluscs known as cephalopods has a history stretching back over hundreds of millions of years. The great disaster that killed off the dinosaurs about 65 million years ago also rendered most species within this class extinct; but some survived and continued to evolve, so today there are around 650 known species of coleoid cephalopods, creatures with eight arms (octopus) or eight arms and two tentacles (squid and cuttlefish). They are found worldwide and vary in size from a few centimetres to many metres. Their brains form a larger fraction of their total mass than is the case for most fish, reptiles and invertebrates.

Many cephalopod species possess a remarkable ability to adjust their colour and appearance in response to their surroundings and circumstances. Some types of octopus can change from pale grey to dark grey in a second or so. Some types of squid have light and dark bands that flow continuously across their body. Such dynamic and reversible changes are due to the presence of cells known as innervated chromatophores. Each such cell can be regarded as a flexible bag containing a pigment. A newly hatched octopus may have only about a hundred such cells, but as it ages the cells increase in number, so that the mature creature may possess more than a million. Around each cell are tiny muscles, arranged radially like spokes around the hub of a wheel. When the muscles are in a relaxed state, the cell adopts an almost spherical form, which minimizes the area of the cell wall. In this state the pigment cells occupy only a small fraction of the total external surface and many of the pigment molecules are concealed behind others. Consequently the overall appearance of the surface is pale. When the muscles contract, each pigment cell is distorted and becomes a thin disc with a diameter that may be more than ten times that of the undistorted sphere. This increases the apparent area more than a hundred times, so that the pigment cells affect most of the total surface, which thereby appears dark. Nerves connect the brain directly to each chromatophore, so that the amount of visible pigment in different areas can be adjusted quickly to match the pattern of the background. This provides a system of dynamic camouflage far more sophisticated than the camouflage of any mammal. Although dark brown pigment is common in the chromatophores, some species have other pigments producing reddish or yellowish colours.

Cephalopod eyes are in some respects more advanced than those of many mammals, with a finer pattern of photoreceptors on the retina and the ability to identify the direction of polarization of the light. Octopus eyes have some features remarkably similar to human eyes, including a flexible lens behind the cornea and a ciliary muscle to adjust its shape. However the curvature of the cornea is much less significant for an eye immersed in water and the eye must rely on the lens to focus the light. To achieve a short and sharp focus, the lens is nearly spherical, and its protein content is distributed so that the refractive index is graded and very high at the centre. This reduces the optical defect known as spherical aberration, and produces a sharp image on the retina. Lens designers have known how to exploit graded refractive indices for less than half a century, but many aquatic creatures have been exploiting such designs for hundreds of millions of years. The giant squid *Architeuthis dux* lives at depths around 500 metres. The largest specimen (found dead) was over twenty metres in length and had eyes with a diameter around 40 cm, the largest eyes of any reliably documented creature.

Squids also possess the ability to distract pursuing predators by ejecting dark ink from an internal ink sac through a funnel. In some languages squid are known by words equivalent to 'inkfish', for example *Tintenfisch* in German and *bläckfisk* in Swedish. Although the function of the discharged ink is sometimes described as the aqueous analogue of a smoke screen, the volume of dark water created is usually too small to provide effective screening. A more plausible explanation is that the patch of dark water diverts the attention of the predator from the fleeing squid. This ink has been exploited since Roman times, and was an important material for artists between the 15th and 18th centuries. It was known as sepia, a word still used today to describe pictures with a reddish brown hue, even when the colour has been derived from a synthetic dye with a different composition. Unfortunately, true sepia is bleached by exposure to light, so that original sepia drawings by artists such as Leonardo da Vinci need to be kept in the dark and viewed in dim light only.

7.6 Lighting up for a mate or a meal

Although many living creatures have skins or coats that absorb and reflect light in a manner that provides camouflage or attracts attention, there are some that possess the impressive additional facility of being able to emit light. The phenomenon of bioluminescence is not known among mammals, birds or reptiles, but occurs sporadically amongst insects, fish, and marine invertebrates. The majority of bioluminescent organisms occur in the sea and they are relatively common at depths below 500 metres, where daylight is effectively absent.

The emission of light by living creatures is often (though not always) based on the oxidation of an organic material, a *luciferin*, in the presence of an enzyme (a biological catalyst) known as a *luciferase*. There are a number of luciferins found among the various bioluminescent species, and each one requires a specific luciferase to generate light. This conversion of chemical energy to light is extremely efficient; virtually no heat is produced in the process. The names of these chemicals are derived from the Latin for 'light bearer'. (The name 'Lucifer' has been applied both to a fallen angel and to matches manufactured in Britain about a century ago and immortalized in a line of a First World War song.)

7.6.1 Bioluminescence in insects

A sophisticated exploitation of bioluminescence occurs among insects. There are more than a thousand species of fireflies in the family *Lampyridae* within the order *Coleoptera*. Fireflies are soft-bodied beetles up to 25 mm long with luminescent panels on the rear abdominal segments, supplied with oxygen via an abdominal air-tube or trachea. Although in some species the larvae emit light, generally it is behaviour of the adult insects that is most intriguing. The species *Lampyris noctiluca* found in chalky areas of the British Isles and elsewhere in Europe has the misleading common name of 'glow-worm'. Figure 7.10 (colour plate) shows a wingless adult female, which emits light from its last three segments in order to attract a flying male. In some types of firefly, the nervous system is able to control the chemical reaction with the luciferin so that the insect emits regular flashes. In one common North American firefly species, *Photinus pyralis*, the males flash for about 0.3 second at intervals of 5.5 seconds as they fly around, advertising themselves and seeking responses from potential mates. The emission is strongest around a wavelength of 567 nm, which is in the middle of the visible spectrum and easily observed by the human eye. The females generally are not airborne and respond by flashing an invitation about two seconds after detecting and assessing a passing male. The timing of the flashes varies from species to species and males are normally able to identify responses from females of the same type. In another North American species, *Photinus consimilis*, the females are more responsive to males that have faster-than-average flashing patterns. The best strategy for males with slower flashing patterns is to accompany a fast flasher and wait for females to respond to him.

Fireflies normally do most of their feeding in the larval stage, and those that continue in the adult form usually have a vegetarian diet. Nevertheless, in North America there is a genus of fireflies in which the females exploit bioluminescence to attract meals as well as mates. Females of the genus *Photuris* adapt the characteristics of their flashes to mimic females of the genus *Photinus*, which enables them to entice *Photinus* males to approach and become food instead of fathers.

Another adaptation of the basic flashing cycle is found in an Asian species, in which the males assemble in a bush and synchronize their flashes to create an impressive spectacle. But the response time is insufficiently fast for matching the individual light pulses when very large numbers of fireflies are assembled in hedges many metres long. The result is a sequence of bursts of light that start at one end of the hedge and finish at the other.

In New Zealand there is a glow-worm *Arachnocampa luminosa* that lives in damp dark caves. It does not belong to the firefly family and displays quite different behaviour. The adult form is known as a fungus gnat and displays no bioluminescence. It is the larval form that emits a bluish green light as a means of attracting small insects, which are trapped by sticky threads. The brightness increases with the time elapsed since the last meal, but there are no abrupt changes in light output. After growing to a critical size, the larva is transformed into a pupa, from which emerges the adult gnat. The gnat has no mouth and devotes its brief life to mating and (if female) to egg production.

7.6.2 Bioluminescence in deep-sea fish

The sea absorbs light quite rapidly, particularly at the red end of the spectrum. As the depth increases photosynthesis becomes progressively more difficult, and virtually ceases about 120 metres below the surface. Creatures that live below this depth are usually part of a food chain starting near the surface. An unaided human eye cannot detect sunlight 500 metres below the surface, because it has not been optimized for low intensity blue light. At this depth many fish species have large eyes directed upward rather than forward, suggesting that they are looking at the faint blue background above. A dark patch moving across the background may reveal the next meal. To avoid being eaten, the hatchet fish *Argyropelecus gigas* attempts to conceal itself from predators below by generating a matching bluish light from photophores underneath its body.

At a depth of a kilometre the sun makes no impression at all, so more than half the species of fish at this depth possess a means of generating light. Many of them maintain a colony of bioluminescent bacteria for this purpose. Anglerfish, such as the female *melanocetus johnsoni*, have a modified dorsal spine with the tip inhabited by light-producing bacteria. The tip is held above the mouth but can be moved independently and the light can be switched on and off. This mobile flashing light acts as a lure; anything that comes to investigate is liable to be eaten. A male of this species is much smaller than a female. He attaches himself permanently to a female, acquiring nutriments by merging blood circulations, becoming little more than a sperm-producing appendage.

Other deep-sea fishes maintain bacterial colonies on their head or side. The bacteria themselves emit the light continuously, but the host fish can control the light output with folds of mobile dark skin or by a mechanism similar to that described earlier in the section on squid colour changes. At the London Aquarium, close to Westminster Bridge, you can see blue flashes coming from the faces of flashlight fish *Photolepharon palperbratus*. Most light emitted by deep-sea fish is blue, and the eye response has evolved to match. It therefore appears a good survival strategy to be coloured red rather than blue. Some predators have moved on to the next stage in the piscine arms race by acquiring the ability to generate and see light at longer wavelengths. (Technology has achieved an analogous shift to longer wavelengths only during the last few decades, providing soldiers and security staff with infrared-sensitive equipment for observations at night.)

The black dragonfish *Malacosteus niger* has separate organs for emitting red and blue light. The red light sources are located just beneath the eyes. There, a fluorescent material absorbs blue light and uses the acquired energy to create red light, which then passes through a filter to select the longer wavelengths. The method by which its eyes detect red light is intriguing and quite different from that of human vision. The retinal photosensitivity for long wavelengths is enhanced by the presence of a compound with a molecular structure somewhat similar to chlorophyll, which absorbs both red and blue light.

7.7 More anatomical oddities

Reflective scales are a common feature of fish. The reflectivity is due to stacks of alternating layers of two materials with different refractive indices. Fish scales consist of guanine and cytoplasm, with refractive indices of 1.83 and 1.34 respectively. Each layer has a thickness roughly equal to one quarter of a wavelength of green light, the condition for ensuring that the weak reflections at each of the numerous interfaces combine constructively to produce a strong reflection overall. In the same manner, many types of beetle and dragonfly have carapaces built in layers so as to produce iridescent reflections.

Some nocturnal mammals make use of multiple-layer mirrors for enhancing vision at low light intensities. Cats have a mirror, the *tapetum*, immediately behind the retina. Any light passing through the retina without absorption gets a second chance after being reflected. The reflective stacks in the eyes of a cat exploit the same physical principle as fish scales, but with different materials.

Although most types of eye exploit refraction to focus light upon the retina, there are a few that use reflection from multiple layer mirrors. The scallop *Pecten maximus* is a bivalve mollusc that needs to make rapid life-or-death decisions as to whether to be open or shut. To assess the situation it has many eyes, which probably provide no detailed visual information but detect movements of dark edges in the vicinity. Each eye is about

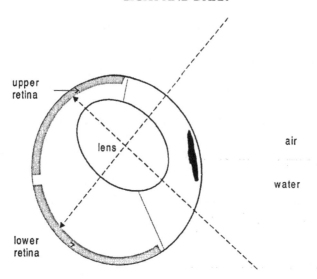

Figure 7.11. An eye for seeing above and below the water surface simultaneously. The eyes of *Anableps* are situated high on the head so that only half of each eye is immersed when the fish swims just below the surface. A dark pigmented strip divides the pupil. A large lens presents different curvatures to light from objects in water and in air, allowing a focus to be achieved on both halves of the retina at the same time.

one millimetre in diameter and incorporates a lens at the front that is too weak to focus the light directly on the retina. However, it has a concave mirror at the back of the eye behind the retina. The light enters the eye and passes through the retina before being reflected and brought to a focus for a second encounter with the retina. Optically, it has much in common with a Schmidt telescope, but evolved hundreds of millions of years before Bernhard Schmidt built the first wide-angle reflector telescope.

A totally different example of eye adaptation is provided by the three species of 'four-eyed' fish of the genus *Anableps*. The Pacific *Anableps dowi* can be found on the coast of Central America. The other two, *Anableps anableps* and *Anableps microlepsis*, occur near the northern coast of South America. Adults of all species are about 25 cm long. Strictly speaking, the fish do not have four eyes, but have two eyes with an external form like a figure of eight. At

148

the aquarium within London Zoo (the one in Regents Park) you can see that these fish often swim at the surface of the water, with the upper parts of their eyes in the air looking upwards and the lower parts in the water looking down. Because air and water have different refractive indices, the eyes have evolved with two parts. Figure 7.11 shows the way the retina and pupil are divided and the way the same internal lens is used to focus the images of objects in and out of the water.

It is not only the eye anatomy of this genus that is remarkable. The fish are viviparous, which means that the females give birth to a few young fish rather than distributing vast numbers of eggs for external fertilization or for meals for other species. Viviparous reproduction is uncommon among fish, although *Anableps* is not the only genus in which it occurs. Copulation for *Anableps* is neither literally nor metaphorically straightforward. The sexual organs of both males and females are asymmetrical, and adopt one of two equally probable sideways configurations that are mirror images. Consequently copulation requires finding a partner not simply of the opposite sex, but also endowed with genitalia configured towards the appropriate side. Human beings should feel grateful that they never encounter this additional problem in their search for a mate.

8

INFORMATION IN LIGHT

8.1 Lighthouses

The idea of guiding sailors by a fixed light has a recorded history that goes back more than two thousand years. One of the seven wonders of the ancient world was the lighthouse of the Pharos, which was on an island close to Alexandria, a city founded in 331 BC by Alexander the Great. Its situation adjacent to the Nile delta enabled Alexandria to develop rapidly as an important trading port and to become a major centre of Greek culture and Arab scientific knowledge. It was an Alexandrian astronomer who provided Julius Caesar with the information needed for reforming the calendar. When the Pharos was completed, around 279 BC, it was the tallest building in the world. It continued to guide ships to the port until its destruction by earthquakes early in the 14th century. Many of the stones were re-used subsequently in the construction of the city's fortifications.

A 12th century Arab visitor recorded a detailed description of the lighthouse, providing much of the information available today about its size and form. It consisted of a substantial base about 30 m square and 56 m high, an octagonal second section and above this a circular tower surmounted by a bronze statue of Poseidon, giving a total height of around 115 m. A central shaft allowed wood to be hauled to the top. During the night a fire burnt at the top, and during the day a mirror reflecting the sun could be seen by sailors many kilometres distant. The legacy of the Pharos includes the words for lighthouse not only in Modern Greek, but also in several Romance languages, such as *il faro* in

Italian. In French *le phare* not only means 'lighthouse' but also 'car headlamp'.

The idea of substantial fixed beacons to aid shipping was developed by the Romans; by 400 AD there were some thirty such structures, including one at Ostia, the port serving Rome. Figure 8.1 (colour plate) shows the lighthouse that was built in 43 AD on the cliff above what is now Dover. Today it is within the grounds of Dover Castle, and has undergone renovation by English Heritage. The structure was modified early in the 15th century for use as a bell tower for the adjacent church, making it difficult to ascertain the exact original height, which was probably between 25 and 30 m. At La Coruña in Spain there is an even more impressive Roman lighthouse over 50 m high, constructed in the 2nd century AD. The Phoenicians traded by sea from their base at the eastern end of the Mediterranean as far as Britain; they also established lighthouses to aid their navigation.

Although Charlemagne was responsible for repairs to the Roman lighthouse tower at Boulogne, there was little development during the so-called Dark Ages, as trade declined. It was not until the 12th century that the construction of lighthouses was resumed in Europe. The *Lanterna* at Genoa was a notable example from this period. With the development of the Hanseatic League the practice spread around the European coast as far as Scandinavia. By the year 1600 there were at least 15 lighthouses along these coastlines.

The 18th century was marked by an acceleration in the development of lighthouses, including towers exposed to the full force of stormy seas. A renowned sequence of remote lighthouses was built on the Eddystone rocks, which lie about 25 km SSW of Plymouth, Devon. The first structure (and its designer) were destroyed by a fierce storm in 1703, only four years after its completion. The second Eddystone lighthouse was largely constructed of wood, and this proved a major factor in its destruction by fire in 1755, after 47 years of service. The third Eddystone lighthouse was designed by John Smeaton, and was built in 1759 using dovetailed granite blocks. It remained in use for over 120 years before the discovery of cracks in the foundations. The upper part of Smeaton's lighthouse was removed and reassembled in Plymouth, where it can still be seen. Its replacement was built on an adjacent rock in 1882 and remains in use today, though with much modernization. Electric lamps have replaced oil lamps and a helicopter pad was added in 1980. Automation has meant that the lighthouse has not been permanently manned since 1982; the need for future visits will be further reduced by the installation of solar cells to provide electrical power.

The first American lighthouse was established on Little Brewster Island at the entrance to Boston harbour in 1716. By the time of the Declaration of Independence in 1776 twelve lighthouses had been completed.

For many centuries burning wood was the source of light. During the 17th and 18th centuries coal gradually replaced wood, but such sources produced copious smoke and too little light. As mentioned earlier, in 1782 Aimé Argand invented an oil-burning lamp that produced very little smoke. (A different Argand invented the diagram for the representation of complex numbers, but they were both Swiss.) The novel feature was a wick in the form of a hollow cylinder, allowing upward flowing air access to the inside of the flame as well as to the outside. Although the fuel for Argand lamps changed from fish oil to vegetable oil and then to mineral oil, the Argand lamp provided the main means of light generation for over a hundred years.

Early in the 20th century the advent of vapourized oil burners and incandescent mantles produced another advance in light generation. In addition, acetylene generated *in situ* from calcium carbide and water was a convenient gaseous fuel, allowing lighthouses to function without constant attention from the keeper.

Carbon arc lamps powered by electricity were invented around the middle of the 19th century. They were used in a lighthouse at Dungeness as early as 1862. In 1913 the Helgoland lighthouse became the most powerful in the world, using carbon arc technology, though oil burning lamps remained the preferred light source until well into the 20th century. Nowadays a wide range of electrically powered sources is available, but the 500 million candlepower light on Ouessant, an island off Brittany, still uses a carbon arc.

Greater visibility at long distances can be achieved by the use of lenses and mirrors to concentrate the light in a required direction, but they need to be rotated to make the lighthouse visible from everywhere on the seaward side. The first revolving beam lighthouse used mirrors, and was brought into service in Sweden

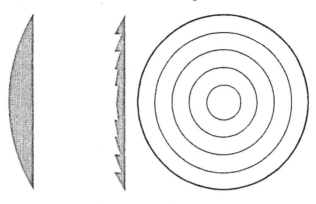

Figure 8.2. Comparison of simple and Fresnel lenses. A simple plano-convex lens (left) has a thick centre. An equivalent Fresnel lens (right) incorporates a series of concentric circular steps, which reduce the thickness at the centre while maintaining the varying angles between the front and back surfaces. In most Fresnel lenses the steps are smaller, closer and more numerous than in this diagram.

in 1781. Because the observer at sea could see the light only intermittently, the ability to produce an identifiable pattern of light and dark with this technology was a bonus. Although at this stage Sweden may have been at the forefront of lighthouse technology, the Swedish language never kept up with developments. Even today a Swedish lighthouse is *en fyr* (a fire) or sometimes *ett fyrtorn* (a firetower) but never *ett ljus* (a light).

Augustin Fresnel was a French scientist who made a number of significant contributions to optics in the early part of the 19th century. These included a rigorous account of the polarization of light. He designed a system of ring prisms to collect light emitted over a wide angle and redirect it in the required direction, avoiding the need for a thick and heavy lens. The modern small Fresnel lens illustrated in figure 8.2 is made in one piece, but the original design, consisting of separate concentric circular prisms, is still used in lighthouses today.

As mentioned earlier, modern lighthouses have their own characteristic sequences of dark and light that provide unambiguous identification. Most patterns belong to one of two large classes. A flashing light is one that is off longer than it is on.

An occulting light is one that is on longer than it is off. A few lighthouses are classified as isophase, meaning that the on and off times are equal. Within the major classes are subdivisions according to the repeating pattern of the flashes or the occultations. The simplest patterns contain single flashes or occultations of constant length at constant intervals. Group sequences contain identical groups that have short intervals within each group and long intervals between successive groups. Composite group sequences have cycles consisting of two or more different sequences, for example alternating between groups containing two and three flashes.

In addition to the temporal variations in the light emitted by lighthouses, there may also be directional variations. These can be used to provide mariners with information about the safest course. Sailors following the recommended route see white light from the lighthouse. If the boat strays too far to port, the light appears red. If the boat strays too far to starboard, the light appears green. This colour scheme matches the colours of the port and starboard lights carried by boats. (A convenient mnemonic for forgetful part-time sailors is supplied by the red colour of port wine.) In south Devon, for example, three-coloured lights with white segments only five or six degrees wide are used to mark the optimum approaches to the harbours at Salcombe (figure 8.3) and Dartmouth.

8.2 Semaphores for optical telegraphy

Historians tell us that fires were used for conveying simple military messages at least three thousand years ago. There is little doubt that relays of fire beacons were used by the Ancient Greeks to send simple messages around 500 BC. Earlier use of such beacons is plausible, but distinguishing fact from fiction can be tricky. The news of the capture of Troy is said to have been conveyed rapidly to Argos by a sequence of burning beacons. However, the first account of this message that can be dated reliably is a play by Aeschylus written about 458 BC, which was several hundred years after the event. The shortest distance from Troy to Argos is around 400 km, which includes some extensive patches of sea. Telescopes had not yet been invented. Furthermore, during

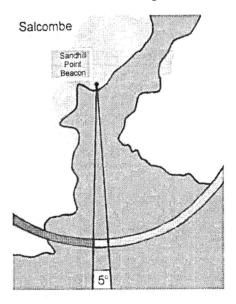

Figure 8.3. Beacon indicating safe route to Salcombe harbour. A sailor on the required narrow course towards the harbour sees the beacon as a white light. If the boat is too far east, the beacon appears green (here represented by the light grey arc); if too far west, it appears red (here represented by the dark grey arc). The beacon can be seen from more than 10 km, which is outside the area of sea shown here.

Odysseus's lengthy journey home after the Trojan War he is reputed to have encountered a giant with one eye in the centre of his forehead. Perhaps this indicates that descriptions of optical arrangements around the time of the Trojan War should be taken with a grain of salt.

However, it is clear from the writings of Polybius in the 2nd century BC that some serious thought about optical signalling had occurred by the time the Roman Empire was developing. (Polybius is usually pronounced with the stress on the second syllable, as in polygamy.) Although Polybius was Greek, he spent many years in Rome and wrote extensive accounts of the city's history. Within these he described a method of coding that allowed texts of military messages to be conveyed by displaying visible signals.

The technique involved the simultaneous display of two groups of identical symbols, such as flags. Each group consisted of between one and five symbols, so that the two groups could provide twenty-five possible combinations. The Greek alphabet contained only twenty-four letters, which allowed a unique combination to represent each letter. It seems that little use was made of this early form of semaphore, probably because the observer needed to be fairly close to identify the symbols displayed.

For thousands of years the use of light signals for sending messages over long distances was limited by the performance of the human eye. By the 13th century it was known that a glass lens in front of an eye could correct for long sight, but it was not until the beginning of the 17th century that it was discovered that two different lenses mounted at opposite ends of a tube could make distant objects appear larger and dim objects more visible. Such slow progress in optics is not surprising because the expounding of scientific theories in Europe was apt to lead to arrest, torture and death, even at the start of the 17th century. However, in the 1660s Louis XIV gave one of his sunny smiles to the Académie des Sciences in Paris, and in the same decade the Royal Society in London obtained its charter from King Charles II. One of the first officers of the Royal Society was Robert Hooke, who made important contributions in many fields of science. He described the use of telescopes in optical communications, and suggested that messages could be transmitted between London and Paris in a matter of minutes via a series of signal towers. However, it was not until a century later that political and military circumstances led to the establishment of long distance telegraphy using light to convey the information.

The pioneer of this method was a Frenchman called Claude Chappe. His career as a priest had been interrupted by the outbreak of the French Revolution in 1789, but the upheaval had some beneficial effects. The new form of government was less reactionary and more willing to consider new ideas. For example the metric system was proposed by the Académie des Sciences in 1793 and approved by the politicians in 1795. Furthermore Ignace Chappe, Claude's elder brother, had manoeuvred himself into a position with significant political clout. Perhaps the most decisive factor in getting support from the government was that France was at war with most of its neighbours simultaneously,

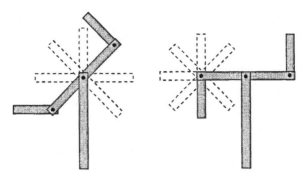

Figure 8.4. Semaphore structure of Chappe's mechanical optical tele-graph. The fixed vertical post supported a pivoted beam with a pivoted indicator arm at each end. The beam was set at one of the four angles shown in the left drawing. Each indicator arm was set at one of the seven angles shown in the right drawing. The eighth angle for which the arm was completely hidden by the beam was not used. The grey drawings show two of the 196 (4 × 7 × 7) possible combinations of the three angles.

which increased the appeal of any proposed new technology of-fering a military benefit.

The first mechanical and optical telegraph link was brought into use in 1794, connecting Paris and Lille, which lie about 200 km apart. Along the route at intervals of around 12 km were relay stations, each consisting of a tower supporting a semaphore structure, illustrated in figure 8.4. The term semaphore implies that the information was represented by the angles of adjustable components. In Chappe's system there were three separately ad-justable angles that could be reset in less than one minute and observed from the next relay station with the aid of a telescope. The fixed vertical post carried a centrally pivoted beam, some-times described as the regulator. At each end of this beam was an indicator arm, pivoted at one end to allow its angle to be set inde-pendently. Ropes and pulleys were used to control the angles. To reduce errors in recognition, the beam was set at one of only four angles and each of the indicators at one of seven. The total num-ber of possible combinations of the three angles was 4 × 7 × 7, which equals 196. This is much larger than the number required for the 26 letters of the alphabet and the ten numerals, so most combinations could represent whole words, thereby compressing the message and allowing it to be sent more quickly.

Some simple arithmetic permits the performance to be compared with more modern telecommunications systems. Because 2^7 equals 128 and 2^8 is 256, each of the 196 distinct settings of the semaphore is equivalent to between seven and eight bits in modern IT jargon. If each successive semaphore position is displayed for one minute, the rate could be described as about 7.5 bits per minute or 0.125 bits per second. (As we shall see in chapter 10, nowadays even 10^{10} bits per second transmitted through an optical fibre does not win any prizes.) A compressed message consisting of twenty successive combinations would take 20 minutes to send to the next relay station 12 km away. Although light travels at almost 300 000 km/s, this complete message effectively travels at only 36 km/h. (This would be the maximum speed at which this message could move from end to end if the entire message needed to be received at one relay station before it was sent to the next.) Although this is slow, it does allow error correction at each relay station. If a small fraction of the combinations is misread, the errors may be obvious and correctable before onward transmission. Without such correction the errors could accumulate as the message travelled on, possibly resulting in gobbledygook for the ultimate recipient. On the other hand there would be no point in delaying the onward transmission if the operators at each relay station were unable to make corrections. Often the messages were enciphered to make them incomprehensible at intermediate stages, because the relay stations could easily be observed by spies. So without waiting for the entire message to arrive, each combination could be sent on from one relay station only two minutes after its arrival. Moving 12 km in two minutes corresponds to a speed of 360 km/h. This is somewhat faster than conveying the message on the TGV, the modern French high-speed train that takes about one hour on the journey between Paris and Lille.

The speed of messages on this mechanical optical telegraph was so much higher than that of any available alternative that an extensive network of links was developed during the next thirty years, in spite of severe turbulence in the French political scene. The second and third links were completed in 1798, connecting Paris to Brest in the west and to Strasbourg in the east. The link to

Lille was extended to Brussels in 1803 and reached Amsterdam in 1811. A transalpine link reached Milan in 1809. (Although these cities are not part of France today, they were under French control in the early years of the 19th century.) By 1823 there were links more than 650 km long to Bayonne on the Atlantic coast near Spain and Toulon on the Mediterranean coast. The French telegraph system employed more than three thousand people in the middle of the 19th century.

It did not take long for the capabilities of the mechanical telegraph to become known in other countries. The second country to adopt mechanical telegraphy was Sweden. Abraham Edelcrantz devised and constructed a system with one intermediate station to convey messages between the royal palace in the centre of Stockholm and the royal country residence at Drottningholm, about twelve kilometres to the west of the city. A royal greetings message was sent by the first version of the telegraph on the birthday of King Gustav IV in November 1794, less than four months after the inauguration of the French link between Paris and Lille.

The Swedish mechanical telegraph differed from the French one in that it did not use a semaphore system. Instead each tower was equipped with an array of ten shutters, each of which could be set either open or closed. In modern IT jargon, each shutter represented one bit and ten bits were sent in each time slot. The ten shutters provided 2^{10} or 1024 combinations. In 1796 Edelcrantz published a treatise on mechanical optical telegraphy that was translated into French and German. The Swedish telegraph system was extended for military purposes, and proved to be particularly valuable during the war with Russia in 1808 and 1809. The technique remained in use in Sweden longer than in any other country in Europe, surviving in the Stockholm archipelago until 1876 and around Gothenburg until 1881.

It might also be said that the French were responsible for the development of mechanical telegraphy in Britain. The Admiralty, based in London, decided that rapid communication with the coast was needed because peaceful co-existence with the French appeared increasingly improbable towards the end of the 18th century. Before the end of 1796 London had been connected to Deal and Portsmouth, each link being about 120 km long with relay stations at intervals of about 15 km. Each relay station normally had a crew of four, two men to observe the adjacent stations

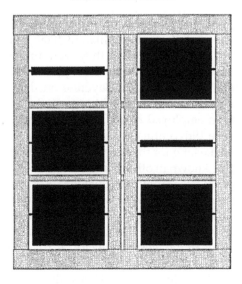

Figure 8.5. Shutter structure for the original British mechanical optical telegraph. The 3 × 2 array of six shutters is shown with four shutters closed and two open. The shutters rotated on a horizontal axis, but the mechanism by which the operators below changed the positions is not shown. The pattern of dark and light here has been arbitrarily selected from 64 (2^6) possible arrangements.

with the aid of telescopes and two to operate the levers controlling the settings. In 1806 the London–Plymouth connection was completed and in 1808 London was linked to Great Yarmouth. The original British system resembled the Swedish one in the use of shutters, which were set either open or closed. They were mounted in a wooden frame about six metres high and could be rotated about a horizontal axis. As shown in figure 8.5, they were almost square in shape and arranged in a three-by-two array, which permits 2^6 or 64 possible combinations. With 36 combinations allocated to letters and numerals there remained 28 for use as shorthand for important and frequently used naval terms. A large model of a British shutter telegraph relay station is displayed at the Royal Signals Museum near Blandford Forum in Dorset, close to the site of one of the former relay stations between Plymouth and London.

When Napoleon relinquished power and was exiled to Elba in 1814, France was no longer considered a threat. In order to save money the British mechanical shutter telegraph links were closed down and dismantled. When Napoleon resumed political and military activities for a few months in 1815 the British telegraph system was in no fit state for conveying to London the news of Napoleon's defeat at the Battle of Waterloo. The information reached Paris much sooner. The ensuing embarrassment in Britain led to renewed enthusiasm for mechanical optical telegraphy. The new British system used neither the same technology nor the same routes as the old one. A substantial post with two semaphore arms replaced the array of shutters at each relay station. Each arm could be set in seven distinct positions at 45° intervals, the two vertical positions (up and down) being treated as indistinguishable. Because the two arms were at different heights they could not be mistaken for each other and so they provided 49 different combinations, equivalent to about 5.5 bits in IT jargon. The link between London and Portsmouth was in use from 1822 to 1847. The route selected lay a little to the northwest of the previous one and had more intermediate relay stations. One of these was restored in 1989 and is now a small museum devoted to the history of this form of telegraphy. Figure 8.6 (colour plate) shows the hexagonal five-storey brick tower at Chatley Heath in Surrey, just over one kilometre southeast of the M25/A3 interchange, and a fifteen-minute walk through the woods from the nearest car park. The pub named 'The Telegraph' in Telegraph Road on Putney Heath in southwest London provides another clue to the route.

Although long-distance mechanical semaphore communication had been discarded well before the end of the 19th century, its little brother lasted well into the twentieth for naval and military communications. This form of semaphore used a pair of hand-held flags, usually with two bright and contrasting colours. The British army used it at the start of the First World War, but so many signallers were killed by enemy snipers that semaphore from the trenches had been abandoned by 1916. However, Lord Baden-Powell, the Chief Scout, continued to regard hand semaphore as a useful skill for members of the scout movement. Competence in semaphore was required to obtain the

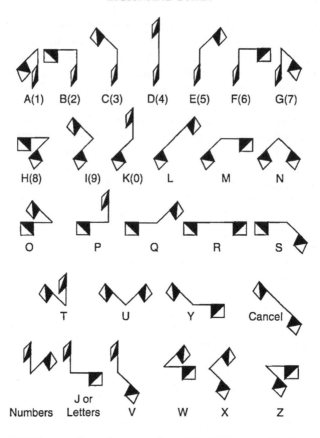

Figure 8.7. Flag positions for semaphore. The 28 flag combinations used by the Royal Navy in the First World War are presented here systematically. They did not exactly match alphabetical order, J and Y being anomalous. Flags were usually coloured red and yellow.

Second Class Badge until the syllabus was radically revised in the 1960s.

The semaphore code used angles at 45° intervals. This provided 28 possible combinations (illustrated in figure 8.7), more than enough for the alphabet. The letters of the alphabet followed a simple sequence, clockwise from the point of view of the observer. For some arcane reason the positions of the letters J and Y were anomalous. The ten numerals had to share symbols with

162

ten letters, though confusion was avoided by using two symbols to indicate a change from letters to numerals or the reverse. Although the basic code had worldwide agreement, the special symbols for error cancellation, start of message and suchlike had local variations.

8.3   Morse, Mance and the heliograph

Although the idea of an electric telegraph first emerged in the 18th century, it was not until the second quarter of the 19th century that much progress was made in this technology. Early electrical telegraphs were constructed by Pavel Shilling in Russia and by William Cooke and Charles Wheatstone in Britain during the 1830s. A major advance was made by the American professor Samuel Morse. He obtained a patent for his idea 1840, but lack of money prevented him from exploiting it until 1843, when he received a grant of $30 000 from the United States Government to develop an electric telegraph line between Washington DC and Baltimore. At each end of the line there was a spring-loaded pen attached to a piece of iron. When current passed through the coil of an electromagnet, the tip of the pen was held in contact with a steadily moving strip of paper. When no current flowed, the spring pulled the pen away from the paper. A switch at each end of the telegraph line was used to control the current through the coil at the other end, enabling a sequence of short or long ink marks (dots and dashes) to be produced on the moving paper strip.

Each of the 26 letters and 10 numerals was represented by a different sequence of dots and dashes. In order to make the Morse code efficient, the long sequences were allocated to the letters that were judged (not entirely accurately) to be the least common. For example, three sequences containing one dot and three dashes were allocated to J, Q, and Y. At the other extreme, a single dot was allocated to E and a single dash to T. Combinations of five dots and dashes were used in a systematic way to represent the numerals. Although the Morse code was devised for messages in English, it can be adapted for other languages with somewhat different alphabets. For example, the German version contains the extra symbols dot-dash-dot-dash, dash-dash-dash-

163

dot and dot-dot-dash-dash to represent the modified vowels Ä, Ö and Ü.

Although the Morse code was devised for electric telegraphy, it was obvious that it could also be used for transmitting messages by light. Oil lamps with movable shutters were used to send messages in Morse code at night during the American Civil War, which lasted from 1861 to 1865. The Royal Navy adopted the method in 1867. Even in 2000 the Royal Navy required its communications technicians to be competent at sending and receiving Morse messages, although the International Maritime Organization abandoned the Morse code for messages of distress in 1999. The highly directional beam from a modern electric signal lamp on a stabilized platform can be used for ship-to-ship messages that are immune to electronic eavesdropping. However it was during the second half of the 19th century, before the advent of radio, that the transmission of Morse messages by light attained its greatest importance. A simple optical instrument known as a heliograph exploited sunlight as the carrier.

When the Indian Mutiny (otherwise known as the First War of Independence) started in 1857, it took around forty days for the news to reach Europe. At that time, sending information to or from India was almost as tedious and slow as sending materials and people. It was not until 1869 that the opening of the Suez Canal reduced the distance that boats had to travel. However in 1863 a young British engineer called Henry Mance (deemed worthy of a knighthood in 1885) was sent to India to work on the installation of electric telegraph cables under the waters of the Persian Gulf and into the Indian subcontinent. After his arrival in India, Mance recognized the need for robust and portable equipment that would provide rapid long-distance communication for soldiers on the move. He realized that reflected sunlight could carry messages in Morse code over long distances, thereby providing an effective method of military signalling in hilly terrain where skies were often cloudless. His first instrument for this purpose was the heliostat, which had a movable shutter to produce the pattern of dots and dashes. In 1869 it was superseded by the heliograph, which achieved the same effect by altering the angle of the mirror. Operation of the heliograph was quicker because it required only a small mechanical movement. Skilled heliograph operators could communicate at ten words per minute over distances that depended on the diameter of the sender's mirror and the quality of the receiver's telescope. Five-inch (125 mm) mirrors were judged to provide the best compromise between performance and portability, though three-inch and ten-inch mirrors were also used. The five-inch mirror could convey messages over 80 km under favourable conditions. The British Army used the method extensively in the Second Afghan War from 1878 to 1880. Military signalling with light reached its peak in the Boer War from 1899 to 1902. Both belligerents made use of the heliograph by day and paraffin-burning lamps by night.

Towards the end of the 19th century in the USA, the military campaigns to wrest land from the North American Indian tribes were in their final stages. Twenty-seven heliograph stations, set up on hills in southern Arizona and New Mexico, contributed to the victory of the American army led by General Nelson Miles over Geronimo and the Apaches in 1886. A reminder of the use of this technology is Heliograph Peak, a mountain about 3050 metres high in the southeast of Arizona. This area enjoys exceptionally clear air, making it ideal not only for heliographs but also for the astronomical telescopes installed later on Mount Graham.

The heliograph was usually mounted on a wooden tripod with adjustable legs, allowing the instrument to be held in a stable horizontal position on uneven or sloping ground. If the Sun was roughly in the same direction as the recipient of the message, one flat circular mirror was sufficient. For other directions two mirrors were needed, arranged as in figure 8.8. A small hole through the main mirror functioned as a backsight for aiming the reflected sunlight towards the receiver. The foresight was mounted on a hinged arm and consisted of a plate with a larger diameter and a larger central hole than the mirror. The operator aligned the two sight-holes with the distant receiver and then adjusted the tilt of the mirrors so that the bright disk of reflected sunlight was centrally located on the foresight. After alignment had been achieved, the hinged foresight could be swung out of the way until needed for a new alignment. The dots and dashes were generated by small changes in the angle of the last mirror, which could be achieved by depressing a knob similar to the spring-loaded keys used in electrical telegraphy.

Marconi's historic demonstration of wireless telegraphy in 1895 and subsequent developments heralded the decline of the

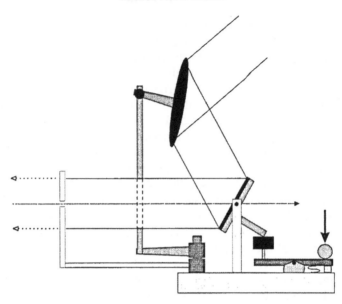

Figure 8.8. Structure of a heliograph. Reflection of sunlight in two flat mirrors produced the broad signal beam, which was aimed towards the recipient with the aid of the foresight on the left and a hole in the centre of the signal mirror. After alignment, the foresight was moved out of the way to allow the full width of the beam to be used. Alteration of the angle of the signal mirror by alternately pressing and releasing the round knob at the end of the cantilever created a message in Morse code. This diagram shows the knob depressed for sending a flash. The instrument was normally mounted on a tripod with adjustable leg lengths.

heliograph. The last wartime use of the heliograph by the British army seems to have been in North Africa in 1941. A few examples are on display at the Royal Signals Museum near Blandford Forum.

8.4 Bell and the photophone

Alexander Graham Bell devised the first electric telephones, though his inventiveness was by no means restricted to one area. He was born in Edinburgh in 1847, one of three sons of Alexander

166

Melville Bell, who was an authority on elocution and methods for teaching deaf people to speak, an occupation that today would be considered a branch of speech therapy. Although Alexander Graham attended a few university lectures after leaving school, much of his training was provided within the family so that he would be able to continue the work of his father. In order to avoid contracting tuberculosis, which had killed both his brothers, the surviving members of the family emigrated to Canada in 1870. The parents remained there; but Alexander Graham moved on to the USA, where he was so successful in applying and teaching his father's techniques that he was appointed Professor of Vocal Physiology at Boston University in 1873. He continued to train teachers for deaf pupils, and studied the physiology and acoustics of the human voice in order to make the techniques more effective.

Bell's interests soon extended to electric telegraphy. Initially he was interested in transmitting several simultaneous telegraph messages through a single link, a remarkable anticipation of modern telecommunication techniques (described in chapter 10). Bell's idea for achieving this involved transmission of sounds of different pitch, each pitch carrying a separate message in Morse code. This may be likened to the ability of a pianist to play different rhythms simultaneously on two or more notes, coupled with the ability of musically competent listeners to pick out and identify the rhythm associated with one of these notes while disregarding the others. Bell therefore turned his mind towards inventing gadgets that could convert patterns of sound into patterns of electrical current or the reverse. Although the original ideas came from Bell, most of the construction of these gadgets was the work of an assistant, Thomas Watson, who was adept at the practical work.

Bell then realized that he could adapt the techniques to transmit a voice. One of the world's most famous patents was granted in the USA on 7th March 1876 (and later in Britain) on 'transmitting vocal sounds telegraphically'. The legendary message requesting Mr Watson's presence was sent a few days afterwards. Later in 1876 Bell demonstrated telephony in an exhibition in Philadelphia, which had been arranged to commemorate the Declaration of Independence in that city one hundred years earlier. (It also coincided with the battle at Little Big Horn in Montana, known as Custer's Last Stand. The battle may have been a disaster for the American army, but it was a boost for the heliograph.) The demonstration in Philadelphia made a considerable impression, perhaps because of rather than in spite of the modest length of the link, which allowed people to walk or run from one end to the other and so convince themselves that the same sounds could be heard at both ends. Early in the following year came public demonstrations of telephone conversations between Bell in Salem and Watson in Boston. This link used an established telegraph line almost 30 km long, temporarily adapted for telephony with microphones and loudspeakers at both ends.

The managers of Western Union, the leading telegraph company, were offered the opportunity to purchase the rights to exploit the invention. They declined, on the grounds that they were unlikely to make any profit from it. A little later they realized that telephones were going to be in great demand, so they entered the business without consideration for the legal implications of the intellectual property. After much wrangling about the patent rights, Western Union was forced to withdraw from the provision of telephones, leaving the field open for development by Bell and the companies he had founded.

It is less widely known that around 1880 Bell and another assistant called Charles Sumner Tainter were granted patents for the photophone, a form of telephony without wires conceived and demonstrated long before wireless (i.e. radio) telephony was pioneered by Guglielmo Marconi. The patterns of the sound waves were conveyed through the air on beams of reflected sunlight instead of on electric currents in copper wires. The transmitter included a mirror that moved in response to a human voice, causing a variation in the intensity of the sunlight directed towards the receiver. Joseph May and Willoughby Smith had discovered in 1873 that selenium conducts electricity more easily when exposed to light, so Bell and Tainter were able to devise a receiver exploiting this photoconductive material to convert the varying light intensity into an equivalent current fed to a loudspeaker. In the first demonstration Bell was at the receiving end and duly waved his hat when asked to do so by Tainter. He wrote later 'I have heard a ray of the Sun laugh and cough and sing.' Although Bell initially regarded this invention as more important than the invention of the electric telephone, the photophone never performed well enough to be a commercial success or to be adopted for military use. Like the heliograph, it would only work during daylight hours in sunny weather; but, unlike the heliograph, the telegraph and the telephone, its maximum range was only a few hundred metres.

It was not until some eighty years later that the use of light to carry the sound patterns of a human voice over long distances began to realize its potential. That is the subject of chapter 10; but it is useful to consider next the developments in semiconductor science and optoelectronics upon which modern optical telecommunications are based.

9

LIGHT IN THE ERA OF ELECTRONICS

9.1 Electronics 1900–1960

9.1.1 Early rectification devices

The seeds of the numerous advances in physics and in electrical engineering during the first half of the 20th century had been sown in the second half of the 19th century. James Clerk Maxwell, a Scottish professor, had produced the equations that describe the behaviour of the entire range of electromagnetic waves, which includes X-rays, visible light and radio waves. In Germany, Heinrich Rudolf Hertz had demonstrated how some kinds of electromagnetic wave could be generated by an electric circuit and used to create electric currents in a separate circuit a few metres distant. Guglielmo Marchese Marconi had developed this technology, and had constructed equipment that transmitted and detected radio signals over increasing distances. By 1898 he had been able to communicate across the English Channel. Actually he may not have regarded this stretch of water as English, as he had an Italian father and an Irish mother. The Italian name *la Manica*, like the French *la Manche*, describes its resemblance to a sleeve rather than its ownership. Intriguingly, the Russian name has been derived by transliteration of *la Manche* into Cyrillic letters, producing a single six-letter word Ламанш that sounds roughly like the French. On the other hand the Swedish name *den engelska kanalen* is a simple translation of the English Channel.

An essential component for radio communications and numerous other electrical technologies is the rectifier, a device permitting an electric current to pass through in one direction only. This component allows the low frequencies of the sound wave to be disentangled from the higher frequency of the radio wave carrying it. The radio wave may be regarded as analogous to a carrier pigeon. Both the wave and the pigeon require methods of detaching the message from the carrier if they are to be useful. Before the end of the 19th century Karl Ferdinand Braun in Germany had devised rectifiers using materials that would now be described as semiconductor crystals. Strangely, his name is rarely mentioned nowadays, although his pioneering work in wireless telegraphy was so highly regarded that he shared with Marconi the Nobel Prize for physics in 1909. In addition to the rectifier he invented the transmitting antenna and the cathode ray oscilloscope, so his Nobel prize was well deserved.

In 1906 Greenleaf Whittier Pickard, working at the American Telephone and Telegraph Company, patented a cheap and simple method of rectification, commonly known as a 'cat's whisker'. This had no feline origin, but was a flexible metal wire with a sharp point. Usually the wire would be made of phosphor bronze, but even an ordinary safety pin would sometimes suffice in an emergency. When the point touched a small piece of silicon or galena (a naturally occurring crystal of lead sulphide), it was sometimes found that the resulting contact would carry electric current in one direction only. Although the cat's whiskers were cheap and simple, finding a suitable position on the crystal surface was a time-consuming trial-and-error procedure.

Even though in the first quarter of the 20th century nobody had yet worked out the underlying physics of junctions between metals and the materials to be known later as semiconductors, their rectifying properties were exploited to make circuits that could receive radio broadcasts. These 'crystal sets' had no local power source and no amplification, but the aerials acquired sufficient power from the broadcast radio waves to create adequate sound in headphones.

The crystal and the cat's whisker were superseded by the thermionic diode valve, which offered a less temperamental method of rectification. The British physicist Owen Willans Richardson was awarded the Nobel Prize for Physics in 1928 for his pioneering work on thermionic emission, which he described in a book with the alluring title 'The Emission of Electricity from Hot Bodies'. In its simplest form, the thermionic diode valve consists of an evacuated glass bulb containing two electrodes, one at room temperature and the other made red-hot by an internal electric heater. Electrons normally have insufficient energy to escape from a cool electrode into the surrounding vacuum, but some electrons do have sufficient thermal energy to escape from a hot electrode. If the hot electrode is made negative (the cathode) and the cool electrode positive (the anode), electrons flow through the vacuum from cathode to anode, but with the opposite electrical bias no electron flow is possible. The name 'valve' was adopted in Britain as electric current could flow in only one direction, making the electrical component analogous to a mechanical valve. In America such a component became known as a vacuum tube or simply a tube, a nomenclature followed in a number of European languages including French (*une tube*), German (*eine Röhre*) and Swedish (*ett rör*), but not Italian (*una valvola*). The Russian term (электронная лампа) is different, being based on the resemblance of the external shape to a light bulb.

A thermionic valve can perform additional functions if it contains three or more electrodes. A third electrode in the form of an open wire mesh or grid, situated between the anode and the cathode, can be given a variable electrical bias that controls the flow of electrons from cathode to anode. Small changes in the grid bias can produce large changes in the current flowing between cathode and anode, so that weak electrical signals fed to the grid are reproduced in amplified form in the output from the anode. Such valves played a leading role in electronic equipment for over fifty years and became increasingly complicated, one type having as many as five concentric grids. The main limitation arose from the hot cathode, which consumed power and heated other parts of the circuit that needed to be cool. Consequently, compact equipment could never be achieved with valves.

9.1.2 The solid-state rectifier

During the first third of the 20th century a number of theories explaining the properties of solids emerged, based on the concepts of quantum physics that had emerged from Albert Einstein,

Werner Heisenberg, Erwin Schrödinger, Max Born, Paul Dirac and Peter Debye. The new theories of the solid state explained some empirical observations about properties of solids that until then had remained a mystery. As early as 1819, Dulong and Petit had noticed that atoms of all elements seemed to have approximately the same heat capacity. In other words, the amount of heat required to raise the temperature of one mole (equivalent to 6×10^{23} atoms) of a solid element by one degree was the same, about 25 joules in the units preferred today. This rule was rather useful in obtaining a rough estimate of the atomic weight of newly discovered elements in the 19th century. Fortunately, nobody had confused the issue in the early stages by finding exceptions to the rule, such as the low heat capacities of common metals at temperatures below $-50\,°C$, or those of hard light elements such as diamond, boron and silicon at room temperature. Well before 1900 Boltzmann had produced an explanation of the basic rule in terms of classical physics, but he could not account for the apparently anomalous low heat capacities. Furthermore, classical physics failed to predict correctly the contribution to the heat capacity from the free electrons that give metals their high conductivity. A satisfactory explanation of the full range of heat capacity data was not available until Einstein applied the new quantum physics and Debye developed this idea further in 1912.

The application of quantum physics to the solid state improved the understanding of many physical properties of solid materials, including their electrical conductivity. Whereas materials in the 19th century were classified as either insulators or conductors, in the 1930s and 1940s there was growing interest in materials with intermediate conductivity known as 'semiconductors'. Their electrical conductivity was very sensitive to the presence of tiny amounts of other elements, so that a wide range of conductivities could be produced in different parts of one crystal if the impurities were not uniformly distributed. Traces of some elements produced a surplus of electrons and traces of others produced a shortage (which implies creating a gap or hole) where an electron ought to have been. Consequently two types of conductivity had been identified, known as n-type when the current is carried by surplus (negative) electrons and as p-type when the current is carried by (positive) holes.

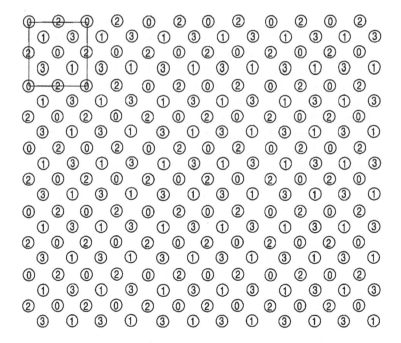

Figure 9.1. Crystal structure of germanium. The crystal is a symmetrical three-dimensional array of germanium atoms, represented by circles. The atoms are situated in different layers (identified by the numbers 0, 1, 2 and 3) parallel to the page. Consequently the unit that repeats to form the three-dimensional pattern has the size shown by the box at the top left. Except at the surface, each germanium atom has four equidistant neighbours. The box size and the layer numbers would be the same in figures 9.3, 9.4 and 9.5, but they are omitted there for simplicity.

The first important target in semiconductor engineering was to replace the rectifying diode valve with a compact and reliable rectifier that did not require a heated cathode. The need for these rectifiers had increased dramatically during the Second World War because radar had become a vital technology. The elements germanium and silicon appeared to be the most promising materials. The arrangement of the atoms within a germanium (or a silicon) crystal is shown in figures 9.1 and 9.2. The two figures correspond to views from different directions. (If you find the layer

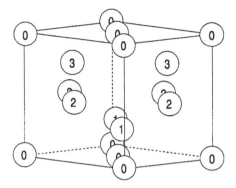

Figure 9.2. Perspective view of the pattern unit in a semiconductor crystal. The arrangement of atoms partially or totally within the box at the upper left of figure 9.1 is shown here. The base of the cube corresponds to the paper on which figure 9.1 is printed and the numbers identifying the layers of atoms are common to both diagrams. The four odd-numbered atoms are entirely within the cube. The even-numbered atoms are situated at the corners or at the centre of the square faces and are shared with adjacent units. The sum of the parts within the cube also comes to four (8 eighths plus 6 halves).

numbers in these two diagrams hard to understand, just ignore them and continue. It is not necessary to absorb all the details of the three-dimensional structure to appreciate the differences between figure 9.1 and later figures illustrating other materials.)

Before the end of the war, a group led by Karl Lark-Horowitz at the Physics Department of Purdue University in the USA succeeded in producing germanium crystals of such quality that robust rectifiers could be made from them. The first germanium rectifiers were based on the properties of junctions between n-type germanium and a metal, but subsequently junctions between n-type and p-type regions of a single piece of germanium were found to be preferable.

9.1.3 The transistor

As semiconductor diodes could replace diode valves for rectification, it was natural to suspect that a new type of semiconductor device with three connections might provide compact and cool replacements for triode valves, which had been used for amplification of electric currents for about thirty years. Shortly after the end of the Second World War, a group led by William Shockley was set up at Bell Laboratories in the USA to investigate ways of achieving this. Just before the end of 1947, they observed amplification in an entirely solid component for the first time. The structure had been devised by John Bardeen and made by Walter Brattain. An n-type germanium crystal grown at Purdue University had been provided with metal contacts in order to investigate the electrical properties of the semiconductor surface and the performance as an amplifier was an unexpected though welcome bonus. The new component was given the name *transfer resistor*, soon shortened to *transistor*. Subsequently Shockley developed the theory of the underlying physical processes. Bardeen, Brattain and Shockley shared the Nobel Prize for physics in 1956.

It became apparent that the amplification process could be achieved in two different but related structures. In the first, the semiconductor was all n-type and the ability of certain types of metal contact to inject holes into the n-type semiconductor was the critical characteristic. Such structures became known as point contact transistors. They were simple to make and dominated transistor development initially. However the manufacture of such components lasted only about ten years because the yield was too low and the waste bins too full. In the second structure, known as a bipolar transistor, there were three regions with different electrical properties (either p, n and p or alternatively n, p and n) within the one piece of semiconductor crystal. Although the production of the bipolar transistors initially posed more problems, they outlasted the point contact transistors and are still being manufactured.

Subsequently another family of transistors, working in a totally different manner, have become prominent. They are known as field effect transistors. As these transistors can be made side by side in vast quantities, they are used in integrated circuits, such as the Intel Pentium chips present in many personal computers. That topic, however, is outside the scope of this book.

9.2 New semiconductors for optoelectronics

Germanium played a leading role in the development of semiconductor technology, and the majority of semiconductor components today are made of silicon, but there are some applications for which these important materials from Group IV of the Periodic Table are unsuited. The limitations are conspicuous in optoelectronics, the subject concerned with the interactions between light and electricity. Although germanium and silicon absorb light and can be used for conversion of light energy to electrical energy, they are of little use for the reverse process, the generation of light from electricity without heating materials to incandescence. For efficient light generation at room temperature and for various additional optical tricks, other semiconductors are essential. The required properties can be found in certain chemical compounds containing two or more elements.

In 1952 Heinrich Welker, a German scientist at the Siemens-Schuckert laboratory in Erlangen, published a classic paper about a family of materials that are known today as III–V semiconductors. These materials are compounds of metallic elements such as aluminium, gallium or indium (which are in group III of the Periodic Table) with non-metallic elements such as nitrogen, phosphorus, arsenic and antimony (in group V). These compounds contain equal numbers of atoms from group III and from group V. In many of the materials the atoms are arranged in a cubic pattern with each atom surrounded by four equidistant atoms from the other group.

Welker found that in several of these semiconductor compounds any surplus electrons move much faster than they do in germanium or silicon. For the first few years after the discovery of these new semiconductors, the interest was centred on their electronic properties, which offered a route towards transistors for high frequencies or for high temperatures. However it was found later that these compound semiconductors possessed other valuable properties that were conspicuously different from those of silicon and germanium. One strange feature is that an increase in the electric field to give electrons a stronger push sometimes has the perverse effect of slowing them down. The explanation is complicated and also unnecessary here, but parents of children two or three years old are familiar with comparable behaviour.

This effect was exploited in a new type of microwave generator, known as a TEO or transferred electron oscillator. However it is the ease with which light and electricity interact in these semiconductors that is relevant to this book.

A thorough explanation of the efficiency of this interaction in many III–V semiconductors is too complex for presentation here. Suffice it to say that it has to do with conservation of momentum. Let us consider a simple but imperfect analogy. When a stunt man impersonating James Bond or some other hero has to leap from one train to another, the jump is easier if the two trains are moving at the same speed in the same direction on adjacent tracks. This rather obvious statement may also be expressed in terms of Isaac Newton's mechanics by saying that the easiest jump involves no change in the momentum of the jumper parallel to the railway track. When electrons in a semiconductor jump from one level to another they change their energy. Energy lost by an electron can be converted into emitted light, while absorbed light can provide an electron with an energy gain. The jumps involving electrons that occur most readily are those for which the change in momentum is zero. Transitions with momentum conservation are easily achieved in many of the III–V semiconductors but not in germanium or silicon.

The most widely used III–V compound is gallium arsenide. The first international scientific conference devoted explicitly and solely to it was held in Reading in 1966. Gallium arsenide does not occur naturally and its name would not be recognized by many passengers on a bus in a traffic jam near Clapham. Nevertheless it has become one of the most thoroughly investigated chemical compounds. It has the simple chemical formula GaAs, which implies that it contains gallium and arsenic atoms in equal numbers. One view of the arrangement of the atoms in a crystal of gallium arsenide is presented in figure 9.3. A similar structure is found with indium phosphide, which has the chemical formula InP and contains indium and phosphorus atoms in a 1:1 ratio.

Another important feature of the family of III–V semiconductors is that all the atoms from each group do not have to be identical, but can be mixtures of two or more elements within the same group. The III–V materials containing three types of atom are classified as *ternary alloys*. The most important of these can be regarded as derived from gallium arsenide by replacing some

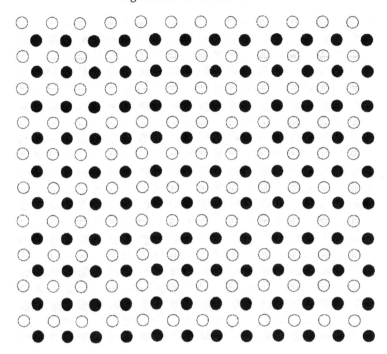

Figure 9.3. Crystal structure of gallium arsenide. The crystal is a three-dimensional array similar to that shown in figure 9.1, except that it contains equal numbers of two types of atom, gallium (grey) and arsenic (black). Gallium atoms occupy the even-numbered positions shown in figure 9.1 and arsenic atoms occupy the odd-numbered positions. Consequently every atom has four equidistant nearest neighbours of the opposite type.

gallium atoms by aluminium atoms. The fraction replaced can be anywhere between 0 and 100 per cent. The chemical formula thus becomes $Al_zGa_{1-z}As$, where z can have any value from 0 to 1. Left to themselves, the gallium and aluminium atoms occupy the available sites within the crystal lattice in a random manner, as illustrated in figure 9.4.

A particularly important feature of the ternary alloys is that the parameter z can be controlled as the crystal is manufactured, so that steps or gradients in the alloy composition can be

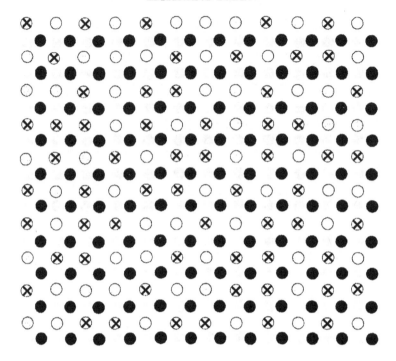

Figure 9.4. Crystal structure of gallium aluminium arsenide. The crystal is a three-dimensional array similar to that shown in figure 9.3, except that aluminium atoms (grey with cross) have replaced half of the gallium atoms (plain grey) in a random rather than a regular manner. The proportions are roughly the same throughout, so the properties of the semiconductor represented here have been modified uniformly by the presence of the aluminium atoms.

incorporated. Because the properties depend on z, it is possible to produce a non-uniform piece of semiconductor in which selected properties have been tailored for a particular purpose. Figure 9.5 shows a simple example of a crystal with both an abrupt and a gradual variation of the composition.

The variable proportion concept can be applied to the group III atoms, to the group V atoms or to both types simultaneously. The III–V materials consisting of four types of atom are classified as *quaternary alloys*. The most important family of quaternary al-

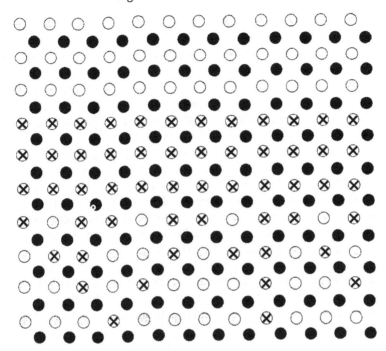

Figure 9.5. Controlled composition variations in a semiconductor alloy. The crystal is a three-dimensional array similar to that shown in figure 9.4, except that the replacement of gallium atoms (plain grey) by aluminium atoms (grey with cross) is no longer random. A large and abrupt composition change can be seen in the upper half of the figure, whereas a gradual change is present in the lower half. Changes in composition can be used to guide light and current in desired directions.

loys can be represented by the formula $In_{1-x}Ga_xAs_yP_{1-y}$ where x and y can be chosen separately and set anywhere between 0 and 1. The resulting availability of a wide range of alloy compositions is a huge benefit for the designer of a semiconductor component, though the people turning the design into a real component may have a more daunting task and more headaches.

The advantages of being able to use non-uniform alloy compositions can be illustrated with a simple analogy. The functioning of components made of an elemental semiconductor such as silicon or germanium (or of a simple binary compound such as gallium arsenide or indium phosphide) can be likened to scoring points at snooker or billiards. Because the surface of the table is perfectly flat, players need considerable skill to persuade a ball to enter a pocket. Incompetent players fail to pot a ball more often than they succeed. The scoring rate would be considerably enhanced if the surface were constructed with slopes designed to assist the balls to move towards the pockets.

In a modern optoelectronic component, spatial variations in the composition of the alloys are deliberately incorporated in all three directions. The variations can be expressed in terms of different values for the abovementioned composition parameters z or x and y in different places within one continuous semiconductor crystal. With a bit of cunning, the designer concocts a structure containing steps and gradients in the alloy composition. The composition profile defines spatial variations in the physical properties of the material. By tailoring these properties, particularly the refractive index and the band-gap of the semiconductor, light and current can be guided along different paths to the required destinations.

9.3 Optoelectronic semiconductor devices

Many optoelectronic components made from III–V semiconductors are classified as diodes, indicating that the piece of semiconductor has two external electrical connections. All the diodes considered in this section contain both n-type and p-type materials, each with an external connection. Most of the important effects occur at or near the junction where the conductivity type changes. As discussed earlier, the original use of p–n junction diodes was rectification, an entirely electrical phenomenon whereby current flows easily in one direction but not in the other. Any p–n junction made of III–V materials, however, can also perform other functions involving both light and electricity, determined primarily by the polarity of the electrical connections.

Multiple functionality is not rare in engineered structures. For example the primary purpose of a table knife is to cut food. However it can also be used to spread butter on bread, push food towards a fork and even turn screws when no screwdriver is

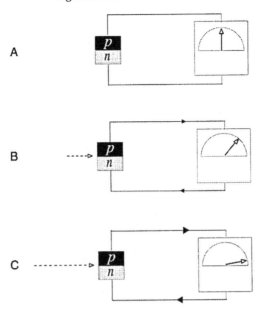

Figure 9.6. The p–n junction creating electrical current from light. This simple electrical circuit contains the semiconductor junction and an ammeter to measure the current. In A the junction is not illuminated, so no current flows. B and C show current generated by illumination within a certain range of wavelengths. At a fixed wavelength, the current increases as the illumination gets stronger. This behaviour may be used to detect light and measure its intensity or to generate current from sunlight.

available. Although a tool or device can function in several ways, it is normally designed for optimum performance in only one or two roles.

Three classes of optoelectronic function are available for exploitation. First, the diode can generate current when exposed to light. Secondly, it can function as an optical modulator, which means that its transparency for certain wavelengths can be changed. Thirdly, it can create light from electric current. These classes of function are illustrated in figures 9.6, 9.7 and 9.8. Each merits being considered in more detail.

183

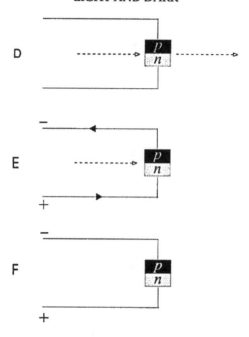

Figure 9.7. The p–n junction as an optical modulator or light shutter. D shows the illuminated junction without any external electrical bias. It is transparent to light within a certain range of wavelengths. E shows the same illuminated junction with a reverse electrical bias, which makes it absorb light in a narrow range of wavelengths previously transmitted and produce an electrical current. F shows the junction with the same reverse electrical bias, but without illumination. The reverse bias alone produces virtually no current, in contrast to the situation with forward bias shown in figure 9.8.

When the p–n junction is neither subject to external electrical bias nor to any other form of energy supply, no current flows. When the diode is illuminated by light of sufficiently short wavelength, the absorbed light provides energy to generate free electrons and holes. In the vicinity of any p–n junction there is an electric field even when no external bias is applied. This field pushes the free electrons and the holes in opposite directions, creating current in any external circuit. As this current is generated by light, it is known as a *photocurrent*. The intensity and wave-

184

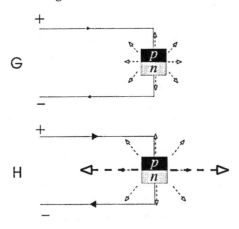

Figure 9.8. The p–n junction creating light from electrical current. When the junction is subject to a forward electrical bias, current readily flows. In some semiconductors, this current generates light at the junction. Light is easily produced by spontaneous emission and emerges in all directions, as shown in G. Under certain conditions, a higher current can also generate light by a second process, known as stimulated emission. In this case the semiconductor performs as a laser and the output is concentrated in particular directions, such as those shown by the two bold arrows in H.

length of the light determine the amount of photocurrent generated. This behaviour can be exploited either to detect light and measure its intensity or to supply electrical power. In practice the design would be optimized for one such application, as shown by the following two examples.

Solar cells (*photovoltaic cells*) generate electric power from sunlight. As explained in chapter 1, the surface temperature of the Sun determines the solar spectrum. The wavelengths with the highest intensity are in the middle of the visible range. Photovoltaic cells based on gallium arsenide and incorporating the ternary alloy mentioned in section 9.2 are suited for inputs in the wavelength range 450 to 850 nm, so they are fairly well matched to the Sun's maximum emission. The surface of solar cells is normally covered by an anti-reflection coating, optimized for wavelengths around 650 nm (red). Consequently the coated solar cells reflect hardly any red and appear bluish violet. Vehicles in space deploy large panels covered with solar cells. Just outside the earth's atmosphere, the total radiation from the sun is about 1.4 kW/m^2. With an efficiency of just over 20 per cent, this implies an electrical output of nearly 300 W for every square metre of solar panel. Although the solar radiation reaching the earth's surface is appreciably less, solar cell panels are also useful at ground level.

For terrestrial applications, sunlight can easily be concentrated by a factor of 100 or more with the aid of a large convex lens or concave mirror, with a proportional decrease in the area of semiconductor required. High light intensities create correspondingly high electrical outputs, so that the metal contacts on the front surface have to be bulky enough to provide a low resistance path for the photocurrent without obstructing more than a small fraction of the incoming light. The intensity of sunlight changes only slowly, and in most cases solar cells are required to provide a steady source of electrical power, so that there is no need for a rapid response to changes in light input. During a visit to Israel, I saw someone displaying one of the most eccentric applications. His hat had a solar cell panel on top and an electric fan attached to the brim at the front, providing a flow of air to cool his face. Appropriately, the fan ran faster in direct sunlight than in the shade.

The detectors of infrared light used in optical communications have to meet quite different requirements. The wavelengths to be detected lie within a fairly narrow range in the infrared, usually between 1300 and 1600 nm, for reasons that will become clear in the next chapter. The optical input may be less than 1 microwatt (10^{-6} W) and localized in a tiny area, so the photocurrents are small and the detectors are required to be very sensitive. They are also need a very fast response as they may have to convert a sequence of light pulses and gaps, each lasting perhaps 100 picoseconds (10^{-10} s) or less, into an equivalent sequence of electrical pulses, without error.

When the p-type material is connected to the negative terminal of an external electrical power supply and the n-type material is connected to the positive terminal, very little current flows in the dark. Illumination at a sufficiently short wavelength creates a photocurrent. One important feature is that the absorption characteristics of the material in the vicinity of the p–n junction are not fixed but depend on the reverse voltage applied, i.e. the transparency of the material can be changed by electrical means. This effect is illustrated in figure 9.7. It can be exploited to produce a component that behaves like an electrically operated camera shutter; but the change between transparent and opaque states can be much more rapid, as no moving parts are involved.

When the p-type material is connected to the positive and the n-type material is connected to the negative terminal of an external electrical power supply, current flows freely, with (positive) holes and (negative) electrons moving in opposite directions from opposite sides, meeting in the thin central layer. As the electrons drop into the holes, annihilating both species, the energy relinquished appears in the form of light if the right kind of semiconductor is used. There are two ways in which such a process can occur. One is known as *spontaneous emission*. As the name suggests, it is not very difficult to produce light by this mechanism but the output tends to go in all possible directions and contains a range of wavelengths. Spontaneous emission is exploited in *light emitting diodes (LEDs)*. On the other hand light can stimulate another process when the concentrations of electrons and holes are sufficiently high. The additional light produced in this way, known as *stimulated emission*, can occur. Lasers exploit stimulated emission to generate a high intensity output in specific directions and within a narrow range of wavelengths. Even a laser produces some spontaneous emission, and stimulated emission does not begin until the current exceeds a threshold value. The two types of output are illustrated in figures 9.8 and 9.9.

9.4 Bright light from cool solids

The principle of stimulated emission was first proposed by Albert Einstein. His theoretical analysis of emission and absorption processes led to the conclusion that the presence of light could stimulate the emission of more light if there were a sufficient population of atoms containing stored energy. The light emitted in this way has the same wavelength, direction and phase as the light that stimulated the process. To allow a high light intensity to be maintained, a laser incorporates a pair of mirrors to reflect light back and forth, with only a small fraction of the light being allowed to escape.

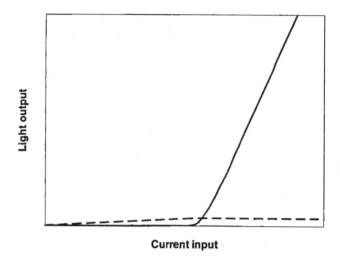

Figure 9.9. Light output from a semiconductor laser diode. Low currents produce only spontaneous emission (shown by the broken line). When the current exceeds a threshold, which depends on the design adopted, stimulated emission (shown by the continuous black line) starts. Thereafter it rises rapidly as the current increases, while the spontaneous emission hardly changes. There are no numbers on either axis because the behaviour is similar in form (though not in magnitude) for a wide variety of semiconductor laser diodes.

We can liken some of the features of stimulated light emission to the outbreak of a craze for a new toy (yo-yos, for example) in a school. For most of the time only a few children show interest in yo-yos. A sudden increase in yo-yo activity cannot occur unless sufficient pocket money is available, but the incentive to own and play with a yo-yo grows dramatically when a substantial number of other pupils are seen playing with them. When the children spend time confined within the school playground, they cannot fail to notice what others are doing. Cash and copying are the keys to the craze.

The first laser was not made from any kind of semiconductor. It was demonstrated in 1960 by Theodore Maiman at the Hughes Research Laboratory in California. The laser light was generated by stimulated emission in a crystal of aluminium oxide doped with chromium (in everyday terms, a rather pale synthetic ruby) between a pair of parallel partial mirrors. The ruby acquired the necessary energy by absorbing green and blue light from an adjacent and powerful electronic flash tube. The output was a narrow beam of red light, at two wavelengths (694.3 and 692.9 nm) so close that the output is sometimes loosely described as monochromatic. In any case, it was totally unlike the mixture of wavelengths emitted by an incandescent source such as the tungsten filament of a GLS lamp. Early lasers were bulky pieces of equipment and were sometimes described as 'a solution in search of a problem', as at that time it was not clear what uses might emerge.

Later lasers have used a variety of materials in various states, including gases such as neon and krypton, liquids containing fluorescent dyes, and solid mixed metal oxides such as neodymium-doped yttrium aluminium garnet. Some are designed for continuous operation at moderate output powers, others to produce extremely short and powerful pulses. They have found many uses in industry and commerce.

In order to achieve a compact laser, alternative and more efficient ways of supplying energy to the piece of emitting material were needed. A direct supply of electrical energy appeared to be possible if the ruby crystal were replaced by a tiny piece of a semiconductor such as gallium arsenide, containing a p–n junction into which electrons and holes could be fed from opposite sides. In 1962 four groups in the USA succeeded in demonstrating infrared lasers functioning in this way. The partial mirrors were neatly provided by the reflectivity of opposite surfaces of the semiconductor crystal itself, so that external parallel mirrors were not needed. Although they were an important step forward, these early diode lasers worked only when cooled by liquid nitrogen to 77 K (−196 °C) and subjected to very short pulses of current separated by long gaps.

It was another seven years before continuous operation at room temperature was achieved. An important factor in this advance was the use of layers of a ternary alloy (discussed in section 9.2) to confine and concentrate the electrons, holes and light inside a thin layer of gallium arsenide. The group at the Ioffe Physicotechnical Institute in Leningrad, led by Zhores Ivanovich Alferov, reported their success in 1970, a few weeks ahead of

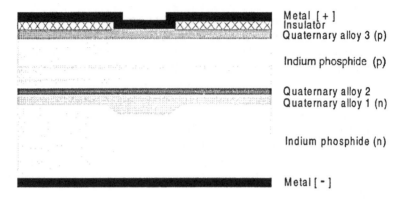

Metal [+]
Insulator
Quaternary alloy 3 (p)

Indium phosphide (p)

Quaternary alloy 2
Quaternary alloy 1 (n)

Indium phosphide (n)

Metal [-]

Figure 9.10. Structure of an IRW laser. This simple laser developed in the early 1980s is a rectangular block with the illustrated cross section extending throughout the entire length. The shapes, dimensions and compositions of the various layers are designed to concentrate at the centre not only the light but also the electrons and holes that carry the electrical current. Consequently intense light emerges from a small area at the centre and travels towards the reader.

Morton Panish and Izuo Hayashi at Bell Laboratories in the USA. Alferov was awarded a share of the Nobel Prize for Physics in 2000, some thirty years later. (In Russian, Alferov is pronounced with the middle vowel stressed and resembling the first two letters in *yonder*, though it is rare to hear an English speaker adopting this pronunciation.)

Since the early 1970s there have been many improvements in the performance of semiconductor lasers. The output power, efficiency and reliability have increased dramatically. At the same time the threshold current for the onset of stimulated emission has fallen. Furthermore, the emission can be set at a precise and stable wavelength within a wide range in the visible and near infrared parts of the spectrum. For reasons to be presented in chapter 10, optical communication systems now require light sources emitting at wavelengths longer than can be obtained from lasers based on gallium arsenide. As a result, such lasers are based on indium phosphide, and incorporate layers of the quaternary alloy mentioned on page 181 in section 9.2.

A simple example of such a laser is shown in figure 9.10. The inverted rib waveguide (IRW) laser, with an emission wavelength of 1300 nm, was developed in the early 1980s and was used in several underwater optical fibre systems, including the British end of the first transatlantic optical link, described further in chapter 10. More recent lasers convert electrical input to optical output more efficiently, have greater stability of wavelengths and can be switched on and off more rapidly, but the complexity of their structures is beyond the scope of this book. What all such lasers have in common is the presence of numerous regions consisting of quaternary alloys with different compositions. Some layers may be less than a hundred atoms thick. The designs take into consideration not only the electrical and optical properties, but also the thermal and mechanical characteristics of each material.

Although the power output from most semiconductor lasers is relatively modest, typically somewhere between a milliwatt and a watt, the light is developed in only a minute fraction of the volume of the semiconductor chip and emerges from a very tiny area. A light output of 10 mW from an area of 1 square μm implies a power density of a million watts per square cm ($10\,000$ MW/m^2). It may be helpful to translate these numbers into a form that is easier to grasp. The surface of the sun emits around 63 MW/m^2, so the light from the microscopic emitting spot on an ordinary semiconductor laser could be described as being about 160 times as intense as the surface of the sun.

Whereas the sun has a surface temperature around 5800 K and much higher internal temperatures, the material in an operating semiconductor laser is only a few degrees above the ambient temperature. The active region has an extremely small volume and the component is designed to have paths with high thermal conductivity for heat removal. Mechanical stress and strain are also important in determining the performance, so that the design of semiconductor lasers involves several branches of engineering.

The first semiconductor lasers had active regions consisting of gallium arsenide and operated in the near infrared at wavelengths around 850 nm. Subsequently, other materials have extended the range of wavelengths considerably. As we shall see in the next chapter, modern optical communication systems use longer, infrared wavelengths, typically between 1530 and 1560 nm. Red lasers operating in the range 630 to 670 nm are now commonplace and used in barcode reading, DVD players and pointers. They are also becoming used in amateur holography. The technology of semiconductors that have nitrogen as the group V element has improved dramatically during the last ten years. The ternary alloy indium gallium nitride has provided access to much shorter wavelengths. Blue semiconductor lasers have become readily available and the Nichia Corporation in Japan has announced that violet laser diodes emitting at 405 nm will be manufactured in large quantities in 2002.

10

OPTICAL
COMMUNICATION TODAY

10.1 Waveguides and optical fibres

The end of chapter 8 contained the story of the photophone, an invention patented by Alexander Graham Bell but only produced in small numbers. It suffered from two major limitations. First, the carrier for the acoustic waveform was sunlight, a commodity unavailable at night and during bad weather. Secondly, the light travelled in straight lines through the air, so that the beam had to be aimed accurately at the receiver and could not penetrate smoke, fog or heavy rain. In contrast, the telephone utilized an electric current, which could be sent along flexible copper cables regardless of the weather or the time of day. Bright reliable lamps and cables capable of guiding light would have made the photophone available twenty-four hours per day.

A waveguide is something that leads waves along a pre-established path. Like a railway track, it functions well when straight or slightly curved but fails if the bend is too sharp. Junctions where one waveguide splits into two are also possible. Waveguides exist for many types of waves. Speaking tubes were used to direct sound waves before electrical telephony became widely available. Sound waves can also be guided by a piece of taut string between two empty tin cans, a form of cheap entertainment and simple engineering enjoyed by many children for more than a century.

As mentioned briefly in chapter 1, the origin of waveguides for light was as much artistic as technological. During the early

1840s the Swiss physicist Daniel Colladon and the French physicist Jacques Babinet independently designed new forms of illuminated fountains in which light was guided along a jet of water that was continuous but not perfectly straight. The guiding of the light within the water was due to the phenomenon of total internal reflection, which occurs when a light ray travelling in one medium (water in this case) falls at a sufficiently shallow angle on an interface with another medium (air in this case). Such fountains remained popular attractions for the next sixty years or more and were showpieces at international exhibitions, including those in London, Manchester and Glasgow in the 1880s.

These early optical waveguides gave rise to the concept of light becoming a commodity like water, gas or electricity that could be supplied from a central source to many distant consumers. In 1880 William Wheeler was granted a patent relating to light distribution through pipes, but it turned out to have negligible commercial value. Before the end of the 19th century the principle of total internal reflection began to be exploited in curved glass rods to provide improved illumination for surgery. Nevertheless it was not until the 1950s that the full potential of optical waveguides in medical procedures was realized. In 1954 Abraham van Heel at the Technical University of Delft in the Netherlands introduced an important advance in the technology. Instead of having the same composition of glass throughout, he produced guides with a core of high refractive index glass surrounded by a cladding with a low refractive index. In these structures, the guiding of light was no longer entirely dependent on internal reflection at the external surface because internal reflection occurred instead at the interface between core and cladding. Consequently the ability to guide light was maintained regardless of the state of the outer surface of the fibre and the properties of the surrounding medium.

In the 1960s the idea of using light as a carrier of telephone messages re-emerged. It was not immediately obvious what would be the optimum method for guiding the light to its destination without losing too much on the way as a result of absorption and scattering. One idea for avoiding loss of light was that the light should travel inside a hollow cylindrical pipe with a highly reflective coating on the inner surface. This technique was investigated for several years but was abandoned in the early 1970s as uncompetitive. The more fruitful approach turned out to be the transmission of light within long glass fibres.

10.2   The transparency of glass

Normally we look through glass that is only a few millimetres thick and consider it transparent for all colours. Viewed edge on, however, a sheet of ordinary window glass looks green and generally murky. Looking through a block one metre thick, we may typically observe that between 20 and 80 per cent of the light is lost in transit, implying that only between 80 and 20 per cent is transmitted. Optical engineers use the term *attenuation* to describe the fraction of light lost in transit, and measure it in decibels (abbreviated to dB), units that are also widely used in electronic and acoustic engineering. A loss of 1 dB means that 79.4 per cent of the input emerges. To get an idea of the dB scale in everyday terms, think of sunglasses. Typical sunglasses might provide an attenuation of 7 dB, which means that the factor 0.794 is applied seven times. As $(0.794)^7$ is about 0.2, we can say that 20 per cent of the light is transmitted.

In the world of commerce and fashion, retailers find the dB scale too complicated and so use a small set of broad categories to describe their wares. Category 1 sunglasses transmit between 80 and 43 per cent and are fairly rare. Category 2 sunglasses may have transmittances anywhere between 43 and 18 per cent, a range that covers attenuations from 3.7 to 7.5 dB. Category 3 sunglasses transmit less light than Category 2 and are the most common type. Category 4 sunglasses are so dark that they are deemed unsuitable for driving. For actually seeing things, these sunglasses are transferred to the top of the head, but the wearer is extremely alluring in spite of bruised shins. It is worth noting that a transmittance of 18 per cent (a 7.5 dB attenuation and the boundary between Categories 2 and 3) is often judged subjectively to be halfway between transparent and opaque.

Comparative newcomers to the field are photochromic lenses, which may be made of glass or plastics. Many of them incorporate silver halides, substances that dissociate and darken on exposure to light – a process also exploited in photography. In the case of photochromic lenses, the darkening occurs on exposure

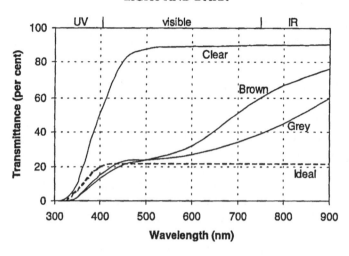

Figure 10.1. Transmission characteristics of photochromic lenses. After a long time in the dark, a typical photochromic material transmits visible wavelengths, but absorbs some violet light and most of the UV. Exposure to short wavelength light alters the material's properties and so reduces transmittance across the entire visible range. Ideally the reduction would be identical for all visible wavelengths, as shown by the broken line, although this is hard to achieve in practice. Materials that appear grey are closer to ideal than those that appear brown.

to UV or violet light. (Tungsten filament lamps, however bright, have hardly any output at such short wavelengths and so have virtually no effect on photochromic lenses.) The reverse reaction occurs continuously, with recovery more rapid as the temperature of the material rises. An ideal photochromic material would transmit uniformly at all visible wavelengths at any level of darkening, absorb all UV and respond rapidly to changes in illumination. Figures 10.1 and 10.2 show that commercially available materials fall short of this ideal. Whereas the iris of the human eye responds in a fraction of a second to changes in illumination, the response of photochromic materials is measured in minutes. Lightening is slower than darkening and its rate depends on temperature. Temperature also determines the degree of darkness achieved during continuous exposure to sunlight. Skiers and snowboarders find their photochromic glasses turn very dark, but

196

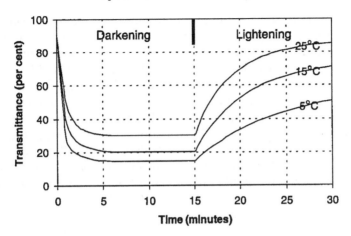

Figure 10.2. Darkening and bleaching in a photochromic lens. Lens material kept in the dark before time zero has a high transmittance, typically about 90 per cent for most of the visible wavelengths. Exposure to sunlight (here for 15 minutes at constant intensity) causes the material to darken, reducing the transmittance. After a few minutes of exposure, it stabilizes at a lower value, which depends on the temperature as well as on the intensity of the sunlight. When the exposure ceases the transmittance is slowly restored at a rate that depends on the temperature.

under a midday tropical sun the same glasses might attain a temperature of 50 °C and their transmission might never fall below 50 per cent. Nevertheless they always do a useful job in removing the UV.

While some scientists have devoted their time to inventing glasses that transmit less light, others have sought ways of making glass ever more transparent. Up to 1960 there had been little interest in transmitting light through glass for more than a few metres, so attenuation of a few dB per metre had been tolerable. The thickest windows in existence were probably those made of lead-containing glass in cabinets for handling highly radioactive materials. In the nuclear industry the loss of half the light in a glass window up to two metres thick is an acceptable price to pay for the absorption of almost all the harmful nuclear radiation.

For optical telecommunications, however, the light needs to travel for much greater distances. In a waveguide made of the best window glass with an attenuation of 1 dB per metre, every ten metres would reduce the light intensity by 10 dB, which translates to a tenth of its previous value. A modest path of one hundred metres through such a glass would mean a reduction in intensity by 100 dB, equivalent to 0.1 applied ten times in succession, or a factor of 10^{-10}. If that figure is hard to grasp, imagine having a very long nose and observing the world through fourteen pairs of 7 dB sunglasses. Obviously glass needed to become much more transparent to permit optical fibres to become useful for telecommunications over many kilometres.

10.3 Optical fibres

In the mid 1960s, Charles Kao and George Hockham, at Standard Telecommunication Laboratories in Harlow, addressed the questions of whether glass could ever be made sufficiently transparent and what impact the availability of such glass would have. The main barrier to progress was judged to be the presence of impurities, particularly iron, in all the glasses available at the time. In 1966 they presented the prediction that glass fibres with much greater purity would be feasible, so that the attenuation could be decreased from 1000 dB/km to below 20 dB/km (roughly equivalent to one set of sunglasses every 350 metres). They also demonstrated that a single glass fibre could be made to carry telecommunications traffic equivalent to 200 000 simultaneous telephone conversations.

Inspired by the possibility of a vast increase in the capacity of telecommunication links, a number of research and development laboratories assembled teams to investigate methods for making optical fibres so low in iron and other transition metal impurities that they achieved the low attenuations needed. In 1970 that Robert Maurer's team at Corning Glass Works in the USA announced the first fibre with attenuation below 20 dB per kilometre. At the red end of the visible spectrum the attenuation was only 17 dB per kilometre. The claim was so important commercially that Maurer needed independent measurements in order to confirm the attenuation. He brought some fibre to Britain, where

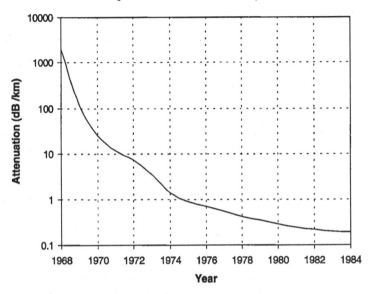

Figure 10.3. Improvement in optical fibre. Purer glass and the use of longer wavelengths led to a remarkable decrease in the light lost as it travelled along optical fibres. By 1970 it was established that optical fibres would become useful for transmission of telephone calls. Only ten years later the glass was so transparent that the first transatlantic optical fibre link was being planned.

(according to the legend) he declined a good lunch in order to stay in the laboratory and guard against any samples of the fibre being surreptitiously retained for subsequent examination and chemical analysis. The precise structure, composition and method of manufacture were closely guarded secrets initially and a rich source of income for patent attorneys during the ensuing decade. The fibre had a cylindrical core consisting of silica with traces of titanium dioxide to increase the refractive index slightly. Around the core was a cladding of pure silica. The trick in the manufacture was that the cladding material was prepared first in tube form. Then the core material was deposited on the inner surface of the tube to create a glass cylinder. This was later softened by heating and greatly elongated by pulling to form the fibre.

Figure 10.3 shows that progress in reducing fibre attenuation did not stop at or even near 20 dB/km. Germanium dioxide became the preferred additive to raise the refractive index of the fibre core. Once the presence of iron had been virtually eliminated, other causes of attenuation became significant. Any hydrogen bonded to oxygen produces absorption bands in the spectrum, particularly at wavelengths around 1400 nm, so steps were taken to eliminate residual hydrogen. Ultimately, however, it is the properties of the uncontaminated material that limit the performance. A significant contribution to the loss of light in any fibre is Rayleigh scattering, the phenomenon also responsible for the blue colour of the sky. As mentioned in chapter 5, Rayleigh scattering is greater for short wavelengths, so shifting from the visible part of the spectrum into the infrared results in lower attenuation. However beyond 1600 nm there is another intrinsic mechanism than dominates the loss. As can be seen in figure 10.4, with silica-based fibres the attenuation is low around 1300 nm and even less around 1550 nm. The manufacturing improvements and optimization of the wavelength reduced attenuation from 20 to 0.2 dB per kilometre in less than a decade. With the latter value, the attenuation of infrared light travelling through 35 kilometres of optical fibre is 7 dB, about the same as the effect of one pair of typical sunglasses on visible light.

The structure of an optical fibre for long distance communication is illustrated in figure 10.5. At the centre there is a narrow cylindrical core, normally 9 μm in diameter, which is less than ten times the operating wavelength. Surrounding the core is the much bulkier cladding. Whereas the cladding glass is pure silicon dioxide, the core composition is silicon dioxide with a small amount of germanium dioxide, which increases the refractive index by almost 1 per cent. Outside the cladding is a protective plastic coating. Individual fibres or bundles of them are encased in tough sheaths to form cables that can withstand rough handling. Undersea cables also incorporate steel wires to provide tensile strength and copper conductors to supply electrical power to remote repeaters.

This type of optical fibre is known as a single-mode fibre because the narrow core and the small step in refractive index at the core-to-cladding interface ensure that there is only one mode of propagation of the infrared light. In this single mode the light

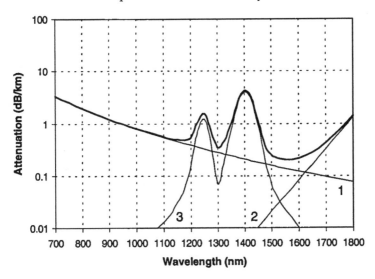

Figure 10.4. Wavelength dependence of attenuation in optical fibre. The diagram covers a range of infrared wavelengths, the red end of the visible spectrum lying at the left edge. Line 1 shows the loss due to Rayleigh scattering. Line 2 shows the absorption by the silicon to oxygen bonds in the silica glass. Line 3 shows two absorption peaks due to the presence of traces of hydrogen bonded to oxygen. The heavy black line shows the total attenuation of infrared light in high quality optical fibres. The lowest loss occurs at wavelengths around 1550 nm.

intensity is greatest along the centre of the optical fibre and is virtually zero at the interface between the cladding and the plastic coating. The light retains its central position even when the fibre is bent rather than straight.

When sounds or data are to be conveyed by pulses of light over long distances it is better for the light to be restricted to a single mode of propagation because multiple modes permit a greater variety of transmission speeds. Nevertheless for short distances, say up to a hundred metres, multimode fibres with larger core diameters (typically 50 μm) are often an acceptable alternative. The concept of a mode is rather too complex for inclusion here, but a crude analogy from the field of transport engineering may be helpful. In a one-way street so narrow that it contains

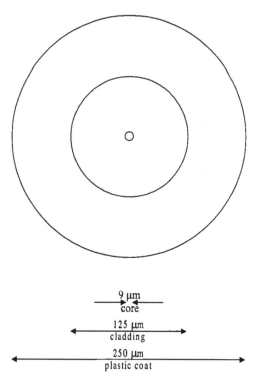

9 μm
core

125 μm
cladding

250 μm
plastic coat

Figure 10.5. Cross section of a typical single-mode optical fibre. At the centre of the fibre is a fine glass core, which has a slightly higher refractive index than the glass cladding that surrounds it. The fibre design ensures that light propagating along the length of the fibre is most intense at the centre and does not interact with the plastic coat, which provides protection against scratches. Because of the material properties and the small diameter, the fibre is flexible and can be wrapped around a finger without breaking.

only one traffic lane, it is impossible to park, overtake or drive on a zigzag course. Vehicles entering this street must emerge in the same order as they enter and their transit times do not vary much. If the street is many times wider, vehicles can progress simultaneously at different speeds and follow zigzag courses that are longer than the direct route along the centre line. Consequently wide streets make it hard to predict the transit times and the order in which vehicles will eventually emerge.

10.4   Optical amplification

As shown in figure 10.3, attenuation of only 0.2 dB per km in an optical fibre was achieved in the 1980s. It remains an impressive achievement, but nevertheless the intensity of an optical signal is reduced by 20 dB after 100 kilometres, i.e. the output after this distance is only 1 per cent of the intensity at the input. The optical signal and its form are still easily detectable at this reduced intensity, but Rome is about 1100 km from Paris and the distance across the Atlantic is almost 6000 km. To ensure that the infrared signal reaches a distant destination with sufficient strength, some form of amplification en route is needed to compensate for the loss.

In early long-distance optical fibre links, the amplification had to be achieved by an optical-electrical-optical technique, known as regenerative amplification. At intervals along the optical fibre a repeater unit was installed. This unit used a photodiode to convert the incoming weak optical signal to an electrical one. The electrical signal was tidied and amplified and then fed to a semiconductor laser to create a new optical signal with a form replicating the original signal, but much stronger. The output from the laser was fed into the next stretch of optical fibre. With regenerative amplification the output wavelength from the repeater is fixed by the semiconductor laser and need not be identical with the incoming wavelength.

A more straightforward method of amplification was devised and demonstrated in 1987 by David Payne and his colleagues at the University of Southampton. They showed that infrared light in the preferred wavelength range 1530 to 1560 nm can be amplified directly inside special optical fibres, without the need for conversion from light to electrical current and back again. The amplification of the light employs the same physical principle as a laser, so that the additional light created has exactly the same wavelength as the light that stimulates the process. The amplification takes place in a short length of an optical fibre with its core composition modified by the presence of some erbium oxide. Erbium is a lanthanide metal with few other uses. The erbium atoms in the fibre core acquire the energy needed to create additional light by absorbing infrared light with specific shorter wavelengths, which is continuously and locally introduced into the special fibre.

The erbium-doped fibre amplifier (or EDFA for short) combines simplicity with versatility. It is capable of amplifying simultaneously many optical signals at different wavelengths within its range. These characteristics make the conversion and upgrading of systems much simpler than when regenerative amplification is used.

10.5   Conveying sound by light

Although light and sound may both be described as waves, there are major differences in their characteristics, such as velocity, frequency and the way in which they propagate. Consequently eyes and ears have greatly different structures. There are various ways in which information about the form of a sound wave may be converted for conveyance by light. The dominant technique in modern telecommunications is quite different from the method used in the photophone described in chapter 8. In 1938 Alec Harley Reeves conceived the principle of pulse code modulation (PCM), a method of representing a sound wave as a sequence of binary digits (ones and zeros). His idea is not limited to one type of carrier and can be applied to electrical currents and radio as well as to light – in fact it is the basis of digital radio and TV and of CD and DVD coding.

Figure 10.6 shows a short section of a sound wave. At regular time intervals the instantaneous amplitude is measured and recorded not as a decimal number but in a binary digital form, i.e. as a sequence of ones and zeros. To make the illustration easier to understand, figure 10.6 shows each measured amplitude recorded as a binary number consisting of only three digits, which means that the amplitude has to be approximated to one of only eight (2^3) values. These digits can be transmitted as an optical signal, in which a pulse of infrared corresponds to a 1 and the absence of a pulse corresponds to a 0. At the receiving end, the process is reversed to reconstruct the form of the sound wave. The reconstruction procedure bears some resemblance to a join-the-dots exercise to create a recognizable shape out of discrete sample values.

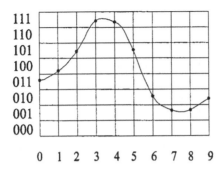

Figure 10.6. Conversion of sound wave to digital format. The curve shows a small part of an arbitrary sound wave. The black dots show samples at regular time intervals. Each dot lies in one of eight horizontal bands, so that its height can be described approximately by a three-digit binary number. Consequently this part of the sound wave can be described by a sequence of thirty bits: 011100101111111101010001001010. This sequence becomes more comprehensible for a human reader if spaces are added: 011 100 101 111 111 101 010 001 001 010.

The use of this very simple procedure produces figure 10.7, which bears some resemblance to figure 10.6, but is far from being a perfect match, because of the rather coarse grid. The procedure is easier to understand with only eight horizontal bands, each identified by a three-digit binary number, but a finer grid is needed to describe the original sound wave more precisely, thereby ensuring that the reconstructed form of the sound wave closely matches the original. The rate of sampling is important too. In order for the signal to be correctly reconstructed, the sampling rate must be at least twice the highest frequency present in the original signal. (This is called the Nyquist criterion, after its discoverer.)

In the realm of contemporary telecommunications, the binary numbers contain eight digits rather than three, thereby providing 256 (which is 2^8) distinct bands. Groups of eight digits are in widespread use in information technology, being known as bytes. Unlike the eight bands shown in figure 10.6, the bands are deliberately allocated unequal breadth. (Here breadth refers to the direction up and down the page.) With 256 bands of increasing

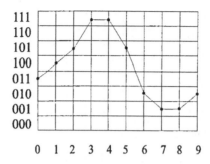

Figure 10.7. Conversion of digital data to sound wave. The black dots are in the centre of the bands that correspond to the thirty-digit bit sequence in the caption of figure 10.6. The data can be used to reconstruct the part of a sound wave shown there. The reconstructed and original versions resemble each other roughly, even in this simplified illustration. In practice a better resemblance is achieved by using a finer grid (characterized by eight-digit numbers), more frequent sampling and a less crude method of connecting the points so that a smooth curve is produced.

breadth instead of eight equal ones, the digital method can represent both loud and soft sounds with high accuracy. The pattern of the sound wave is usually sampled 8000 times per second and eight binary digits describe each point, so the form of the wave is represented by a sequence of 8 × 8000 binary digits per second. This rate may be expressed as 64 000 bits per second or 64 kbit/s. The sampling rate of 8000 times per second is sufficient to convey sound frequencies up to 4000 Hz. The absence of higher frequencies may make hi-fi enthusiasts wince, but the frequency range is adequate for clear speech communication.

If the sequence describing one sound wave is being sent at 64 kbit/s, the maximum time available to send each bit is 1/64 000 s, which comes to about 16 μs. Although this may seem rather a short time, it is extremely long compared with the time required for switching laser light from on to off or *vice versa*. This means that the pulse of infrared signifying that a bit is a 1 and not a 0 could be made much shorter than the 16 μs available before the following bit requires attention – leaving a time gap that can usefully be filled.

One of the benefits of digital optical communication is that it readily permits multiplexing and demultiplexing. These procedures resemble the kind of trick in which a conjuror fills a pint bottle with a pint of beer and a pint of milk simultaneously, passes the bottle around the audience and then extracts the two drinks again in an unmixed state. I cannot say how that conjuring trick is performed, but figure 10.8 presents a simple example of mixing and unmixing on a printed page. The figure shows the way information identifying two words can be compressed into the space required for only one word and then recovered. This method of reducing paper consumption is unlikely to become widespread, though a related technique is used on the little cards that present different images from different viewpoints. In the realm of telecommunications multiplexing and demultiplexing are widely employed in order to maximize the traffic in each optical fibre.

If the indication of a 0 or a 1 can be achieved in half the maximum time available, the gaps between the bits are long enough to be used for a second bit sequence representing another sound wave. Figure 10.9 illustrates how two sequences of three digit binary numbers can be interleaved to convey along the same route the bits representing two unrelated sounds. By making each bit even briefer, the number of independent sequences that can be interleaved increases. The technique of interleaving sequences in this way is known as *time division multiplexing (TDM)*. Each bit can now be sent in such a short time that it is routine to send 10^{10} bits per second (10 Gbit/s) through one optical fibre. This corresponds to over 150 000 simultaneous telephone conversations in a single fibre, though not all the capacity can be used in this way, because some must be devoted to control and supervision of the system.

Naturally there is a limit to how brief a bit can be, because it is essential that each bit remains distinguishable from the bits before and after it. Whereas in empty space all wavelengths travel at the same velocity, this is not necessarily so in a glass fibre. If a pulse of infrared light contains a range of wavelengths, its progress along the fibre may not be at a single velocity. In this case, the pulse becomes dispersed (longer but less intense) as it progresses. This effect may be likened to despatching groups of runners on a long track at regular intervals. If the runners within a group progress at unequal speeds, the group will become more widely spread as

Figure 10.8. Multiplexed words on a printed page. The two words on the top line remain unmistakable on the second line, where half of the lettering has been obscured by a regular pattern of white bars. If the white gaps within each word on the second line are filled by the black parts of the other word, both words can be written in the space previously needed for one word, as shown on the third line. Although the third line is difficult to decipher by the unaided human eye, either word can be made obvious by selectively obscuring the other word. The fourth line is derived from the third by superimposing a regular pattern of grey bars in two different positions.

the distance run increases. If the starting intervals are too small for the length of the course, the gaps eventually will disappear and the groups will overlap.

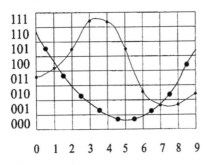

Figure 10.9. Interleaving of two sound waves in digital format. In addition to the curve shown in figure 10.6, this figure contains a curve representing a small part of a second sound wave. The larger black dots show regular samples of the second wave, allowing it to be represented by the sequence of three-digit binary numbers, **101 011 010 001 000 000 001 010 100**, shown here in bold type. (As in the caption of figure 10.6, the spaces are only for the benefit of the human reader.) The sequences for the two waves can be interleaved to form a single sequence describing both waves: 01**1**10**1**10**0**01**1**10**1**01**0**11**0**01**1**11**0**00**1**01**0**00**0**10**0**10**0**01**0**10**0**01**1**00**0**10.

However there is a simple method that allows more groups of runners to use the track simultaneously without confusion: provide clothes of different colour for groups that might overlap. Analogously, infrared pulses at several different wavelengths allow a big increase in the number of bits per second that can be conveyed simultaneously in an optical fibre. This multiple wavelength technique is called *wavelength division multiplexing (WDM)* and can be used in addition to TDM. The ability of modern optical components to separate and identify the wavelengths is remarkable, and may be likened to the recognition of several hundred distinct colours within the visible spectrum. A set of 32 wavelengths at intervals of 0.8 nm fits easily inside the wavelength range 1530 to 1560 nm, the region of low attenuation shown in figure 10.4. With each wavelength operating at 10^{10} bits per second, the total throughput becomes 32×10^{10} or 0.32×10^{12} bits per second. This is equivalent to five million simultaneous telephone conversations in a single optical fibre. With TDM and WDM together, a single optical fibre could provide sufficient capacity for the entire population of Denmark to make an international telephone call simultaneously.

10.6 The long and the short of optical communication

Considering that Bell's invention of the telephone was in 1876, ten years after Brunel's *Great Eastern* had succeeded in laying a telegraph cable from Ireland to Newfoundland, it may seem rather strange that there were no telephone cables of any type crossing the Atlantic Ocean until 1956, the year before the launch of Sputnik 1 by the USSR. Up to that time transatlantic messages were sent either as a Telex text or as speech by short wave radio. The first transatlantic telephone cable connected Scotland and Newfoundland. It used copper wires, with thermionic valves for amplification, transistors then being regarded as too unreliable for such an important link. It could cope with just thirty-five simultaneous conversations. During the next thirty years several more transatlantic copper cables were installed, increasing the total capacity to a little more than ten thousand simultaneous conversations.

A major change in the strategy was adopted at the start of the 1980s, with the decision that the eighth transatlantic telephone link (known as TAT 8, for reasons that are not hard to guess) should be based on optical fibres instead of electrical conductors. The operating wavelength was set at 1300 nm in the infrared. At this wavelength the dispersion in the available optical fibres was close to zero. In simpler terms, small variations in the operating wavelength would have virtually no effect on the velocity of the infrared light in the fibre. TAT 8 required most of the decade to plan, construct, test and lay on the ocean bed. Consequently it was not until 1988 that this system came into service, connecting the USA with two countries in Europe. It had been considered politically expedient to have a Y-junction south of Ireland and to bring the eastern end of the cable ashore in both England and France. The immediate effect of TAT 8 was a quadrupling of the capacity of the underwater telephone links between Europe and North America. The effect on the total transatlantic capacity was less dramatic, as radiotelephony via satellites had become available in the 1960s with the launch of Telstar in 1962 and Early Bird, the first geostationary satellite link, in 1965. As mentioned in chapter 4, a geostationary satellite orbits the earth at the same angular rate as the earth rotates, thereby maintaining a constant position above a chosen place on the equator.

Nevertheless, cable routes have the advantage of being much more direct. The distance across the Atlantic Ocean near 45° N is less than 6000 km. The radius of the orbit of a geostationary satellite is fixed by the mass of the earth and exceeds 42 000 km. As the radius of the earth is almost 6400 km, the transatlantic radio signal via a geostationary satellite has to travel about 72 000 km, which takes about a quarter of a second or twelve times as long as the signal in an underwater optical fibre. Although such a delay may seem trivial at first glance, the difference between a cable link and a satellite link can be noticed in a simple transatlantic telephone conversation, and becomes significant when two computers are interacting.

After the success with TAT 8 more and better optical fibre links were installed across the oceans. As the output of the tiny semiconductor lasers was becoming more stable and fibres could be made with zero dispersion at longer wavelengths, the operating wavelength moved from 1300 to 1550 nm. Within another ten years the total number of speech channels available across the Atlantic had reached the equivalent of one and a half million, though by this stage additional uses (videoconferences and pornographic pictures on the internet are two examples) had been found for the facilities available.

With optical fibres able to handle so many bits per second it is not surprising that much of the long distance telecommunications traffic today is conveyed in the form of pulses of infrared light in optical fibres. The number of optical fibre cables across the Pacific Ocean reached double figures some years ago. The map of the optical fibre cables crossing the Mediterranean Sea resembles a pile of spaghetti. Even the Channel Islands and the Isle of Man have optical fibre cables connecting them to the British mainland. Of course, optical fibre links are not confined to the sea bed and the majority of optical fibre cables are installed across land. Nowadays most of the traffic consists of data being transferred from computer to computer, telephone conversations forming only a minor part of the total. Optical fibre technology

211

provides such a cheap and reliable service that organizations such as international airlines often send enquiries thousands of miles to wherever labour costs are low rather than employ staff in the country of the caller.

Although optical fibres are the preferred method for conveying high-density telecommunications traffic over long distances, they are often not cost effective at the ends of the link. The connection between a local exchange and a private house is still usually in the form of a copper cable rather than an optical fibre. This often remains the case even when the subscriber has paid for additional facilities such ADSL (asymmetrical digital subscriber line). ADSL brings a digital signal all the way to the subscriber's premises, and is capable of carrying several simultaneous telephone conversations or a one-way video input.

Mobile telephones connect the user to the service provider's network in a quite different way. They use radio signals, but the right to exclusive use of a narrow band of radio frequencies now costs so much that suppliers of third generation mobile telephones find themselves in deep financial difficulties. On the other hand nobody is required to pay for the right to use particular optical wavelengths. As a result there has been renewed interest during the last few years in the use of light to carry digital signals over the first or last few kilometres of a telecommunication link. The light could be guided along optical fibres, but there are also a number of small companies developing systems in which the light is aimed directly at the receiver and travels through the air. Rapid installation is possible, because no cable is needed. The line-of-sight transmission concept has been resurrected decades after the demise of the heliograph, though the new technology is much more sophisticated than that developed by Mance in the 19th century. A precise single wavelength in the red or infrared is preferred to the mixed visible wavelengths in sunlight; optoelectronic gadgets function hundreds of millions of times faster than army signallers and the code is very different from that of Morse. However, free space optical communication outdoors still works best in fine weather!

BIBLIOGRAPHY

Booth K M and Hill S L 1998 *The Essence of Optoelectronics*, (Englewood Cliffs, NJ: Prentice-Hall)

Bowers B 1998 *Lengthening the Day – a History of Lighting Technology* (Oxford: Oxford University Press)

Deligeorges S 1999 Les obscures zébrures du zèbre *La Recherche* **316** 42–3

Duncan D E 1998 *The Calendar* (London: Fourth Estate)

Farrant P A 1997 *Colour in Nature* (London: Blandford)

Greenler R 1989 *Rainbows, Halos and Glories* (Cambridge: Cambridge University Press)

——1994 Sunlight, ice crystal and sky archaeology *The Candle Revisited* ed P Day and R Catlow (Oxford: Oxford University Press) ch 4

Harrington P 1997 *Eclipse! The What, Where and How Guide to Watching Solar and Lunar Eclipses* (New York: Wiley)

Hecht J 1999 *City of Light – the Story of Fibre Optics* (Oxford: Oxford University Press)

——1999 *Understanding Fibre Optics* (Englewood Cliffs, NJ: Prentice-Hall)

Heilbron J 1998 Les églises, instruments de science *La Recherche* **307** 78–83

Jones T 2000 *Splitting the Second – The Story of Atomic Time* (Bristol: Institute of Physics Publishing)

Land M 1988 The optics of animal eyes *Contemp. Phys.* **29** 435–55

Littmann M, Wilcox K and Espenak F 1999 *Totality – Eclipses of the Sun* (Oxford: Oxford University Press)

Ninio J 1998 *La Science des Illusions* (Paris: Éditions Odile Jacob)

Park D 1997 *The Fire within the Eye – a Historical Essay on the Nature and Meaning of Light* (Princeton, NJ: Princeton University Press)

Perkowitz S 1996 *Empire of Light – a History of Discovery in Science and Art* (Washington, DC: Joseph Henry Press)

Porter V 1996 *Goats of the World* (Ipswich: Farming Press)

Pritchard D C 1990 *Lighting* (Harlow: Longman Scientific and Technical)

Rice T 2000 *Deep Ocean* (London: The Natural History Museum)

Smith D 1995 *Optoelectronic Devices* (Englewood Cliffs, NJ: Prentice-Hall)

Solymar L 1999 *Getting the Message* (Oxford: Oxford University Press)

The San Diego Evening Tribune July 31 1937; article on *Charles Sumner Tainter* reproduced in
http://history.edu/gen/recording/ar304.html
Tovée M J 1995 Les gènes de la vision des couleurs *La Recherche* **272** 26–33

INDEX

accommodation, 103, 106, 107
achromatopsia, 114
ADSL, 212
Aeschylus, 154
agouti hair, 139, 140
Alexander's dark space, 96
Alexandria, 67, 150
Alferov, Zhores, 189
Anableps, 148, 149
analemma, 32
anglerfish, 146
anomalistic month, 62
arc lamps, 9, 15
Argand, Aimé, 7, 152
astrologers, 22, 65
atmospheric effects, 33–35, 84–102
atomic clock, 39
attenuation, 195–203
Australian time zones, 78, 79
axial rotation, 41

Babinet, Jacques, 194
Babylonians, 22, 37, 75
Bacon, Roger, 69
bacterial light, 146
badger, 134
Bardeen, John, 174
Bay of Fundy, 65
bees, 7, 131
Bell, Alexander Graham, 166–169, 193, 210

belted Galloway cow, 136, 137
Benham's disc, 123–125
bent line illusions, 119, 120
Berthollet, Claude, 16
bioluminescence, 144–147
bird colours, 130, 133
blind spot, 105, 106
blue tits, 130
Bologna, 71
Brattain, Walter, 176
Braun, Karl, 171
Brewster angle, 89, 97
Bunsen, Robert, 8

caesium, spin-flip transition, 39
café wall illusion, 119, 121
calendars, 67–76
candles, 6, 7
Caroline Island, 82
Caroline Islands, 114
Cassini, Gian, 71
cataract, 106
cats, 138–141
cat's whisker, 171
cave paintings, 6
cephalopods, 142–144
cerium dioxide, 9
Chaldaeans, 55
Chappe brothers, 156
Chatley Heath, 161

checker board distortion,
120–122
chiaroscuro art, 15, 16
chiasma, 118
Chinese calendar, 75
Chinese time, 78
chlorophyll, 1, 2
chromatophores, 142
chromosomes, 113, 130,
139–141
CIE overcast sky, 98, 99
ciliary muscle, 103, 104, 143
cinematography, 17, 116
Clement IV, Pope, 69
clouds, 98
coal gas, 8
Colladon, Daniel, 194
colour blindness, 112–114
colour uniformity
distortion, 123
colour vision, animal,
129–131
colour vision, human,
109–114
Concorde, 63
cones, 108–115, 129–131
conservation of momentum,
178
Co-ordinated Universal
Time, 40
Copernicus, 71
cornea, 103, 104

Dalton, John, 112
Danti, Ignazio, 71
da Vinci, Leonardo, 15, 143
dawn of new millennium,
81–83
day, definitions, 26, 27
day, length variations, 29–31

day, starting time, 26, 38
daylight, 31–36
Debye, Peter, 173
declination, 22
deep oceans, 2, 144, 146
detection of light, 183
deuteronopia, 113
digital optical signals,
204–210
diodes, 174, 182
Dorset, 65, 66, 160
Dover, 65, 66, 151
dragonfish, 147
Dulong and Petit rule, 173

Earth, orbital eccentricity,
25, 26, 29
Earth, rotational variations,
38, 39
Earth, tilt of axis, 29
earthworms, 4
Easter, 75–76
eccentricity, 22–25, 43, 49
eclipses, 55–63
eclipse year, 59
Eddystone, 151
Edelcrantz, Abraham, 159
EDFA, 204
Edison, Thomas, 9
Einstein, Albert, 172, 173,
187
electricity from light,
183–186
electric lighting, 9–15
emission spectra, 3, 14
English channel, 170
enlightenment, 5
Ephemeris Time, 39
equinoxes, 19–22, 25, 29, 32,
33, 36, 53, 68

eye anatomy, animal, 143,
147–149
eye anatomy, human,
103–109

field of view, 105
fireflies, 144, 145
fireworks, 16
first point of Aries, 22
fiscal year, 73
fish, 133, 146–149
flame colours, 8, 17
flashlight fish, 146
Fleming, Sandford, 77
fluorescent lighting, 14
fovea, 104, 109, 110
French semaphore links,
156–159
Fresnel, Augustin

Galileo, 23, 72
gallium arsenide, 178–179,
182, 189–191
gas discharge lamps, 13
gas lighting, 8, 9
geostationary satellite, 50, 51
germanium, 174–178, 182
Ghiraldi, Luigi, 69
glass, 195–202
glow-worm, 144, 145
GLS lamp, 10–12, 189
goats, Swiss, 136–138
Golden Number, 75
gravitational fields, 64, 65
Greenwich Mean Time, 26,
29, 37, 38, 77–82
Gregorian calendar, 21, 24,
40, 67, 69–70, 73–75
Gregory XIII, Pope, 69
Gulf of Mexico, 65

Gurney, Sir Goldsworthy, 8

Haidinger, Wilhelm, 114
Halley, Edmund, 56
halos, 99–102
Handel, 15
Harvest Moon, 53, 54
Hayashi, Izuo, 190
heliograph, 164 -166, 168
Helios, 5
Hermann's grid, 123, 124
Hertz, Heinrich, 170
Hipparchos, 19
Hockham, George, 198
holes, 173
hotel lighting, 10, 11
Hoyle, Fred, 46

ice crystals, 99–102
Iceland, 35
Iceland spar, 91
Ides of March, 68
indium phosphide, 178, 182,
190
Inquisition, 72
International Atomic Time,
39
International Date Line, 58,
62, 77, 80–83
inverted rib waveguide
laser, 190, 191
iodopsin, 109
iris, 103, 104

Jewish calendar, 73–75
John Paul II, Pope, 72
Julian calendar, 68–70, 73
Julius Caesar, 67, 68
Jupiter, 65, 72

Kao, Charles, 198

Kepler, Johannes, 23
Kepler's laws, 23–25
Kiribati, 81, 82

Land, Edwin, 91
Lark-Horowitz, Karl, 175
lasers, 62, 187–192
lateral geniculate nuclei, 118
Lavoisier, Antoine, 7
leaf colour, 1, 2
leap seconds, 40
leap years, 68
LED, 187
lemur, 134, 135
lens of eye, 103–107, 148
LGN, 118, 119
light bulbs, 10–13
lighthouses, 150–154
limelight, 8
Line Islands, 82
line-of-sight
    telecommunications,
    212
local noon variation, 31
longest day, 29
longest daylight, 33
luciferin, 144
Lumière brothers, 17
lunar calendars, 73–76
lunar day, 48–50
lunar nodes, 43–48
lunar orbit, 41–43, 46, 47
lunation, 42
Luther, Martin, 69, 72

Maiman, Theodore, 188
Mance, Henry, 164
mantis shrimp, 131
Marconi, Guglielmo, 165,
    170, 171

Martian atmosphere, 84
Maurer, Robert, 198
Maxwell, James Clerk, 170
Mayan calendar, 76
mercury vapour lamp, 14
meridian, 22, 71, 77
Metonic cycle, 74, 75
midnight Sun, 35
Mie scattering, 86, 98
Millenium Island, 81–83
modes, 200–202
modulation of light, 184,
    186, 187
mole rats, 4
monkeys, 129, 130
month, anomalistic, 62
month, calendar, 67, 68,
    72–75
month, sidereal, 41–43, 74
month, synodic, 42, 48, 74
Moon, 5, 41–66
moonlight, 26, 41, 50–54, 108
moonrise and moonset,
    44–46, 51–53
Morse, Samuel, 163
Morse code, 163–164
Müller-Lyer illusion, 119
multiplexing, 207–210
Muslim calendar, 73

Nepalese time, 80
Newgrange, 4
Newton, Isaac, 25, 63
Niaux cave paintings, 6
Nicea, Council of, 68, 69
Nichia Corporation, 192
Nobel prizewinners, 2, 171,
    175, 190
Nyquist criterion, 205

occulting light, 154

October Revolution, 67, 70
octopus, 142, 143
Old Testament, 5, 6
op art, 126–128
optical amplification, 203
optical fibres, 198–204, 210–212
optical illusions, 118–128
optical telegraphy, 156–161
orbit of the Earth, 24, 29, 71, 72
orbit of the Moon, 41–43
Orthodox Church, 70, 76
ozone, 3

panda, giant, 134–136
Panish, Morton, 190
parhelion, 102
Payne, David, 203
Pharos, 150
phosphors, 112
photochromic lenses, 195, 196
photocurrent, 184
photopic vision, 109
photophone, 166–169, 193
photoreceptors, 105, 107–114, 129–131
photosynthesis, 1, 2
Pickard, Greenleaf, 171
Pingelap, 114
Pitt Island, 81–83
Planck, Max, 2
planetary orbits, 23–25
polarized light, 88–94, 97, 114, 115
Polaroid, 91, 93, 97, 114, 115
Pole Star, 20
Polybius, 155

precession of the equinoxes, 19–21
protanopia, 113
Protestants, 69, 70, 72
Pulfrich's pendulum, 116, 117
pulse code modulation, 204
pupil, 103, 104
pyrotechnics, 16

quantum physics, 172
quaternary alloys, 180, 190

rainbows, 94–97
ravens, 130
Rayleigh scattering, 84, 85, 91, 92, 200, 201
rectifiers, 170–175
Reeves, Alec, 204
refractive index, 33, 94, 101, 103, 182, 194, 200
relative distances of Sun, Earth and Moon, 61, 64
relative masses of Sun, Earth and Moon, 43, 64
religious views and influences, 5, 23, 69–72
retina, 104 -108, 115
rhodopsin, 107, 108
Richardson, Owen, 171
right ascension, 22
Riley, Bridget, 128
road vehicle lighting, 87
Robert-Houd, Paul, 17
rods, 101, 107–110, 115, 129
Roman Catholic Church, 23, 69–72
Romans, 67, 68, 72, 151, 155

Royal Signals Museum, 160, 166
Royal Society, 156
Russian time zones, 78, 79

St Malo, 65
Salcombe, 154, 155
Saros cycle, 55–62
satellite radiotelephony, 210, 211
satellite visibility, 51
scallop, 147
Scotland, 73
scotopic vision, 107, 108
second, definitions of, 37–40
Selene, 5
semaphore, 156–162
semiconductors, 173–192
sepia, 143
Shakespeare, William, 6
Shockley, William, 176
sidereal day, 27–29
sidereal month, 41–43, 49, 74
sidereal year, 20
signal lamp, 164, 165
silicon, 174, 177, 178, 182
single mode optical fibre, 200–202
Sirius, 41
six pip time signal, 38
sky, 84, 85, 94–99
sodium vapour lamp, 13
solar calendars, 67–70
solar cells, 185, 186
solar day, 27–30
solar radiation, 3, 25
solar transit, 27
solstices, 21, 29, 31–33, 36, 67
son-et-lumière, 17
Sosigenes, 67

South Pole clocks, 80
speed of eye response, 115–118, 123–126
squid, 142, 143
Stonehenge, 5, 45
Sun, 5, 19–36
sunburn, 86, 130
sundial, 30, 31
sundog, 102
sunglasses, 195
sunlight, 2, 3, 164–166, 168, 191, 193
sunrise and sunset, 33–36
Swan, Joseph, 9
Swanage, 11
Sweden, 70, 152, 153, 159
synodic month, 42, 48, 60, 74

TAI, 39, 40
Tainter, Charles, 168
tapir, 136
TDM, 207
telegraphy, 154–166
ternary alloys, 178
thermionic valves, 171
thorium dioxide, 9
tides, 39, 41, 48, 63–66
tilt of Earth's axis, 29, 31, 48
tilt of lunar orbit, 46
time systems, 38–40
time zones, 77–80
Tonga, 80
transatlantic telephone cables, 210, 211
transferred electron oscillator, 178
transistors, 174–176
transparency of glass, 195–201
tritanopia, 113

tropical year, 20–22, 38, 39
Troy, 154
tungsten filaments, 3, 10–13, 189, 196
tungsten halogen lamps, 12, 13
twilight, 84–88

Universal Time, 38, 40
USA, times and dates, 78, 80

Valais blackneck goat, 137, 138
valves, 172
van Heel, Abraham, 194
van Honthorst, Gerrit, 16
Vasarély, Victor, 126
Venus, 24, 25, 65, 72
vision, 103–128, 148
visual acuity, 109
visual cortex, 118, 119
visual purple, 107
vitamin A, 108
von Welsbach, Carl, 8

water jets, 17, 194
Watson, Thomas, 167
waveguides, 193, 194
waxes for candles, 7
WDM, 209
Welker, Heinrich, 177
Wheeler, William, 194
white nights, 78
Wilkes Coast, 81, 82
Wordsworth, William, 94

xenon arc lamps, 15

year, Jewish, 74, 75
year, length, 19–22, 38, 39, 59, 68, 69
year, Muslim, 74
Year of Confusion, 68
year, starting date, 72, 75
Young, Thomas, 112

zebras, 131–133
zebra finch, 133
zebra fish, 133

T - #0372 - 101024 - C10 - 229/152/14 - PB - 9780750308748 - Gloss Lamination